Digital Computer Process Control

Digital Computer Process Control

CECIL L. SMITH
Associate Professor and Chairman
Department of Computer Science
Louisiana State University
Baton Rouge

INTEXT EDUCATIONAL PUBLISHERS
College Division of Intext
Scranton San Francisco Toronto London

The Intext *Series in*

CHEMICAL ENGINEERING

Consulting Editor

PAUL W. MURRILL

Department of Chemical Engineering
Louisiana State University

ISBN 0-7002-2401-7

Library of Congress Catalog Card Number 78-185819

Printed in the United States of America by The Haddon Craftsmen, Inc., Scranton, Pennsylvania.

To my parents

Preface

Where are we headed in computer control? This question has certainly received its fair share of attention, and an additional discourse on the topic seems inappropriate. However, when writing a textbook on the topic, the author must address such questions in order to decide what material to include and what to omit. Since this author's opinion on the subject greatly influenced the selection of topics for the book, a very brief answer seems in order. The word *opinion* should be stressed, since authoritative answers to such questions simply do not exist.

During a panel discussion at the San Francisco American Institute of Chemical Engineers Meeting (December 1971), the growth in the use of computers in process control was compared to the so-called "steamboat" curve. With the introduction of the steam engine in the early 1800s, it was natural to consider its use as a source of power to propel a ship, and a number of steamboats were built. However, they did not become an immediate success, largely because the manufacturers of sailing ships steadily improved their product (for example, the clipper ships were introduced), thereby maintaining a competitive position. In fact, the steamboat's share of the market remained about constant until the late 1800s, when it rather quickly proceeded to capture virtually the entire market.

Are we on a "steamboat" curve with respect to computer control? With the commercialization of the digital computer in the 1950s, its potential for use in process control was quickly pointed out. The first applications appeared in the late 1950s, followed by even more applications in the 1960s. But since the manufacturers of conventional control hardware have steadily improved their product, the net result has been that computer control has not "swept the market."

With the introduction of the minicomputer (defined in any manner you like) in the late 1960s, a wide range of computer hardware became available. This has provided the user with considerable flexibility, which will undoubtedly result in more applications. Coupled with the continuing reduction in the cost of computing power, will this lead to a "sweep of the market"?

This could possibly be true, even assuming the end user continues to practice control in much the same fashion that he does with conventional instrumentation. However, in most respects, the capabilities of the computer far exceed that of conventional control hardware, but this approach makes little use of the real potential for the computer. The fact is that while we have seen great advances in the computer itself, we have seen precious little advancement in what we know about how to utilize its capabilities effectively.

It is the author's opinion that when substantial progress is made in this latter area, then the computer will "sweep the market." Therefore, this text stresses the control theory aspects of computer control. One chapter giving a brief description of computer hardware, software, and interfacing is included. The remaining eight chapters are devoted to a spectrum of control techniques and concepts. The author does not pretend to have all the answers as to how computers should be used in process control, but a wide-enough spectrum is included to acquaint the reader with at least some of the possibilities.

This book arose out of the author's activities in three areas. First, although practical topics are hopefully emphasized, the book has been rather heavily influenced by graduate research and graduate instruction activities. This manuscript has been class-tested in a graduate course on digital control at Louisiana State University for the past two years. Second, the author has participated in more than thirty industrially-oriented short courses on automatic process control and digital process control. Portions of the manuscript have been used in several of these during the last two years. Third, the author has been active as a consultant in process control.

The author is sincerely grateful for the help and encouragement from many sources during the preparation of this manuscript. Special thanks must go to Dr. Paul W. Murrill for his assistance and advice over the years. Thanks must also go to Drs. Donald B. Brewster and Philip H. Emery, who gave the author his first opportunity to really practice computer control. Dr. Armando Corripio has made several suggestions for improvement, and has used the manuscript in a graduate course. Credit also goes to the graduate students and short course attendees for their helpful comments. Sincere appreciation is extended to all those who have worked in the preparation of the manuscript, especially to Mrs. Sonja Hartley for typing and to Miss Nina Stewart for proofreading.

Cecil L. Smith

Baton Rouge, Louisiana
March, 1972

Contents

chapter 3 SUPERVISORY COMPUTER CONTROL . . . 55

chapter 4 MATHEMATICS OF SAMPLED-DATA SYSTEMS . . . 82

chapter 5 FREQUENCY-DOMAIN CONSIDERATIONS . . . 108

chapter 6 CONTROL ALGORITHMS . . . 136

chapter 1

The Role of the Computer

For the Silver Jubilee issue (January 1970) of *Instrumentation Technology*, the Journal of the Instrument Society of America, the readers overwhelmingly elected the digital computer as the most significant development during the previous quarter-century. Furthermore, virtually all the experts contributing to a forecast for the future saw the role of the digital computer expanding in all aspects of process control.

The technology of applying digital computers to process control has developed rapidly since the late 1950's. The hardware limitations and problems that plagued early installations have for the most part been solved. The software maze that surrounded computer applications is being resolved, both by standardization and by replacement of software functions with hardware.

Perhaps the largest gap exists between the available control theory and the control concepts used in the applications. It is in this latter area that this text is primarily concerned, describing approaches by which the concepts of control theory can be incorporated into a practical control system.

However, it would be meaningless to discuss control theory without any mention of software and hardware, since all of these topics are intertwined in a complex manner in the final installation. But do you first discuss hardware and software followed by control theory (i.e., how the computer is used to control a unit)? Or do you discuss control theory followed by hardware and software? Obviously we have a "chicken and egg" type of problem.

The approach taken in this text is something of a compromise.

In this first chapter we shall discuss some of the common relationships between the computer and the process, indicating the role the computer plays in regulating the process. With this basis we shall briefly cover the hardware, software, and computer/process interface for digital computer control systems. The remainder of the text is devoted to control concepts.

Throughout this text the question of primary concern is "How can I use a computer to generate a larger economic return from my process?" In this chapter we shall first examine the process control problem and the conventional approach to process control. Then we shall examine four approaches to applying digital computers in process control: data logging, direct digital control (DDC), supervisory, and hierarchical.

1-1 THE PROCESS CONTROL PROBLEM†

Industrial processes for which successful computer control systems have been reported include blast furnaces, petroleum and petrochemical plants, paper machines, and textile mills. Each has its unique problems, so this discussion must be fairly general. In all of these processes, the variables are divided into the four categories illustrated in Fig. 1-1.

1. *Manipulated variables.* These are variables such as input raw material flow rate, steam pressure in a vessel, etc., whose

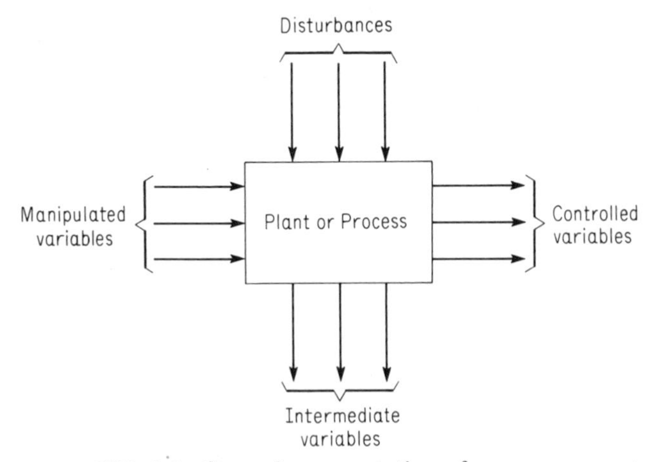

FIG. 1-1. General representation of a process.

†Much of the material in this and subsequent sections of this chapter has been reproduced from C. L. Smith, "Digital Control of Industrial Processes," *Computing Surveys*, Vol. 2, No. 3 (September 1970), pp. 211–241, with permission from the Association for Computing Machinery.

values can be adjusted by the control system, whether analog (conventional) or digital.

2. *Disturbances.* These are variables whose values affect the operation of the process but which are not subject to adjustment by the control system. Examples include composition of raw material and ambient air temperature. Some variables in this category can be measured while others cannot.
3. *Controlled variables.* These are the variables whose values really measure the performance of the plant, and thus are those which the control system must keep at some target value (often called the *set point*). Examples include production rate and product quality. The general control problem is to adjust the manipulated variables so as to maintain the controlled variables at their target values in face of disturbances. Some controlled variables can be measured directly, but some must be inferred from other measurements, a task in which digital computers far excel their analog counterparts.
4. *Intermediates.* These variables appear at some intermediate point in the process. The control system can often use them advantageously in determining what control action should be taken. Examples include temperature of a water jacket and composition of an intermediate stream.

Because a typical plant has several variables in each of the above categories, it is apparent that the control of process units is no simple matter. This is further complicated by the difficulty in deriving a mathematical model of the process from process characteristics. The problem in this regard is that the process characteristics depend first, on the level of plant operation (i.e., the plant is usually highly nonlinear), and second, even at a constant operating level the plant's characteristics change with time (i.e., the plant is nonstationary).

Although complicating the job of installing the computer control system, these aspects are really the basic reasons such a sophisticated control system can be justified. The ability of the digital computer to collect large quantities of data, analyze it, and make logical decisions based upon the results makes it attractive for such applications.

1-2 CONVENTIONAL CONTROL SYSTEMS

Before delving into the characteristics of digital control systems, an appreciation of the conventional approach to process control is a helpful background (1,2).† The basic control loop in conventional

†Numbers indicate References at end of chapters.

(analog) systems is the simple feedback loop illustrated in Fig. 1-2. The value of the controlled variable is detected by a sensor or transmitter (e.g., a thermocouple for measuring temperature). This value is compared to the desired value or set point to generate the error.

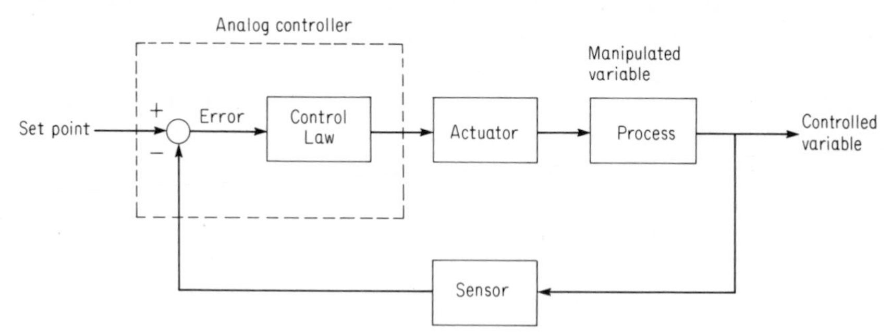

FIG. 1-2. Basic conventional feedback control loop.

The control law generates a change in the manipulated variable so as to drive this error to zero. This controller output is imposed upon the process by an actuator, which is an automatic positioning valve in many cases.

The control law commonly used is the proportional-integral-derivative (PID) relationship or some simplification thereof. That is, the manipulated variable $m(t)$ is related to the error $e(t)$ by the equation

$$m(t) = K_c \left\{ e(t) + \frac{1}{T_i} \int_0^t e(\tau)\, d\tau + T_d \frac{d\, e(t)}{dt} \right\} + m_R \qquad (1\text{-}1)$$

where K_c = proportional gain
T_i = reset or integral time
T_d = derivative time
m_R = reference value at which the control action is initialized

The adjustments K_c, T_i, and T_d appear generally as adjustments on the rear or side of the controller. The selection of their proper values is normally a trial-and-error procedure called "tuning," although some systematic approaches have been suggested (1,3–6). In well over 75 percent of the applications, only the proportional-integral (PI) terms are used, primarily because of the difficulty of tuning the general PID (or three-mode) controller.

In a typical plant there may be anywhere from a few of these devices to upwards of a hundred or more. Until around the late

1950's these devices were invariably pneumatic (operated on air pressure). Aside from being much more reliable than their vacuum-tube electronic counterparts, they had the added advantage of safety when used in areas in which explosive gases might be encountered. Only with the introduction of solid-state electronic controllers in the late 1950's are pneumatic controllers gradually being replaced.

No matter whether pneumatic or electronic, conventional analog control systems basically suffer from inflexibility. There must be almost a one-to-one correspondence between control loop functions and hardware to perform these functions. This places several burdens upon the designer of the control system:

1. His strategy must be such that it can be implemented with analog hardware.
2. Subsequent modifications of the control strategy require modifications of the analog hardware.

In the mid-1950's control-system designers began to look toward the digital computer as a means for overcoming these problems. Virtually any control strategy is programmable, and most modifications in the strategy are simply program changes.

1-3 DATA LOGGERS

As illustrated in Fig. 1-3, the data logger essentially is not directly active in the control or regulation of the process. Instead, it simply records the values of the important process variables at regular intervals of time. The sales pitch for the data logger goes something like this: "From the data generated by logging all pertinent process variables, you will learn so much about your process that the optimum operating strategy will be obvious." Unfortunately life is not this simple. Much of the data will be redundant information taken at or near the normal operating conditions. Process modeling requires

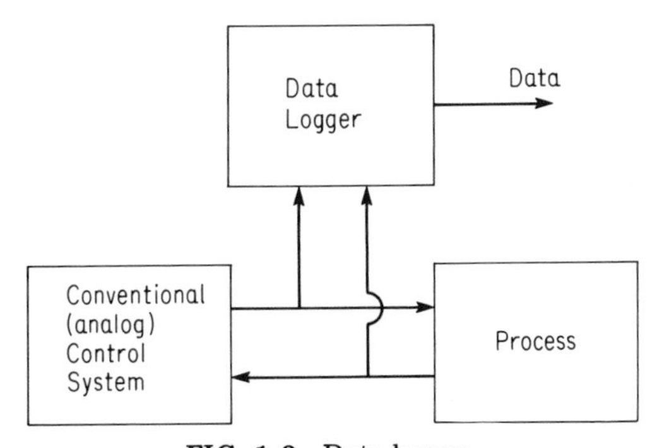

FIG. 1-3. Data logger.

carefully devised process tests to obtain the necessary data. While good data logging is vital during these tests, data logging in itself is not adequate.

Except in special cases, data logging is not sufficient to justify a digital computer. In fact, Wherry and Parsons (7) question the addition of data logging to a process control system in one of the other three categories (direct digital control, supervisory, or hierarchical). Exceptions include nuclear power plants, where the Atomic Energy Commission requires that certain records be maintained, and laboratory automation systems, where data is the primary product.

For the reasons given in the above paragraph, this concludes the consideration of data loggers in this text.

1-4 DIRECT DIGITAL CONTROL (DDC)

In direct digital control (8–11) the computer calculates the values of the manipulated variables (e.g., valve positions) directly from the values of the set points, controlled variables, and other measurements on the process. The decisions of the computer are applied directly to the process, and hence the name *direct digital control* or DDC. This control arrangement is illustrated in Fig. 1-4.

As the values of the manipulated variables are calculated by the computer, the conventional three-mode controllers described above are no longer needed. Their functions are instead performed by the

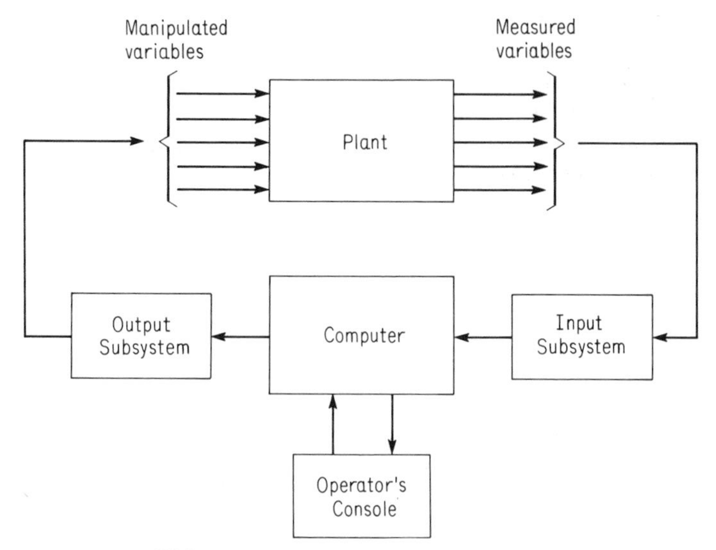

FIG. 1-4. Direct digital control (DDC).

equations, called *algorithms*, by which the computer calculates the manipulated variable from the set point and controlled variable. An example of a control algorithm is the discrete (finite-difference) equivalent to Eq. 1-1 for the continuous controller

$$m_n = K_c e_n + \frac{K_c T}{T_i} \sum_{i=0}^{n} e_i + \frac{T_d K_c}{T} (e_n - e_{n-1}) + m_R \tag{1-2}$$

where m_n = value of manipulated variable at the nth sampling instant
e_n = value of the error at the nth sampling instant
T = sampling time

Other parameters are defined as before. This equation is called the *position form* of the control algorithm, as its result is the actual value of the manipulated variable, typically a valve position. If Eq. 1-2 is written for m_{n-1} and subtracted from Eq. 1-2 as given above, the result is

$$\Delta m_n = K_c(e_n - e_{n-1}) + \frac{K_c T}{T_i} e_n + \frac{T_d K_c}{T} (e_n - 2e_{n-1} + e_{n-2}) \tag{1-3}$$

where $\Delta m_n = m_n - m_{n-1}$ = the change in the manipulated variable, and hence the name *velocity algorithm.* The significant difference between Eqs. 1-2 and 1-3 is that Eq. 1-3, the velocity algorithm, does not contain the term m_R, the value of the manipulated variable when the loop is placed on automatic (i.e., calculations are begun.) If Eq. 1-2 is used, the computer must be able to read the valve position to insure a smooth transition from manual to automatic, called *bumpless transfer.* Using Eq. 1-3 accomplishes the same objective without having to read m_R.

Being the digital equivalent of the three-mode analog controller, the above algorithms are used in by far the majority of the applications. However, there are other alternatives, some of which produce algorithms whose performance exceeds the performance of the above algorithms. Both of these approaches are considered in Chapter 6.

One of the first incentives suggested for DDC systems was economic savings. The basic idea was that since one computer could provide the same functions as several analog controllers, there must be some point at which the cost of these several analog controllers would equal the cost of the digital system. Some early estimates on the number of loops to be replaced ranged as low as 50, but was unfortunately not proven to be correct. Two problems seemed to have been underestimated:

1. *Programming costs.* With no prior experience and without DDC software packages or proven executives, the programming effort far exceeded that anticipated.
2. *Backup hardware.* This problem stems from the fact that operating personnel must be able to exercise effective control over the plant in event of total computer failure. In many cases this backup was a complete analog system, thus eliminating any hardware savings.

In essence, analog control systems are very reliable and relatively inexpensive, the result being that there is no breakeven point at which the cost of the analog control elements equals the cost of the digital system to replace them. There are a few exceptions, one being where a single digital system replaces nearly identical analog systems on several nearly identical units.

Although reduction in manpower is another potential justification for DDC systems, this has rarely proven to be feasible. Most process units are already operating with a minimum of operations personnel. In fact, the presence of a computer often entails the presence of higher-caliber personnel, causing costs to increase rather than decrease.

No matter what the application, whether accounting or process control, the use of a digital computer virtually always increases the cost as compared with the "conventional" approach to accomplishing the task. The justification must come from improved performance of the digital system. In the case of the direct digital control computer, we shall show in Chapter 5 that simply replacing a conventional analog controller with its equivalent discrete control algorithm does not increase the basic performance capabilities of the control system. That is, performing digitally the same tasks that would have been performed with an analog system does not generate significant economic returns, if any. Digital computers are expensive, primarily because of their well-known computational capability. To justify this expenditure of funds, the computational capability must be utilized fully by implementing control strategies that are either impossible or impractical to implement with analog hardware.

One alternative is to use algorithms designed via the z-transform techniques described in Chapter 6. Although the performance of these algorithms exceeds that of the PID algorithm in virtually all cases, the improvement is generally about 30 percent, which is probably not enough to justify a digital system for the typical plant. There are exceptions, one case being systems with large "dead" times, such as a paper machine. Here the improvement is very substantial.

This reasoning leads to the conclusion that in many cases, the best approach is to implement digitally only those control loops in which there is significant improvement in control performance, leaving analog control systems on the other loops. This is the so-called hybrid approach to process control, which became even more attractive with the advent of the minicomputer. In addition to basic feedback schemes, it is attractive to digitally implement adaptive systems, noninteracting control, on-line tuning, dead-time compensation, and other similar approaches covered in Chapter 8. Again, the economic justification must come from the implementation of control strategies that are impossible or impractical to implement with analog hardware.

Although the justification of DDC systems is not easy, there are certain cases where these systems are extremely attractive. For example, the operation of batch reactors involves considerable sequencing and logic, tasks at which digital systems are ideally suited. Similarly, the logical decisions required during startup and shutdown can be implemented digitally much easier than via an analog system.

Parsons, Oglesby, and Smith (12) report the following application of DDC:

> *Process*: Phillips Petroleum Company's 500 million pounds/year ethylene plant at Sweeny, Texas.
> *Total control loops*: 180.
> *Loops on DDC*: 120, leaving 60 on analog. These loops were those that the operations people felt were absolutely necessary to keep the plant operating.
> *Analog backup*: 20 percent for the 120 DDC loops.
> *Computer*: Systems Engineering Labs 810A with 16K words core storage.
> *Reliability*: 13 hours downtime in 32 months.

On the technical side, the system was rated a definite success. If the installation were repeated, many of the 60 analog loops would be placed on DDC.

On the economic side, the picture is not so bright. Little economic return was realized from the DDC portion of the system. Most of the return was from supervisory functions which were implemented after the initial startup. The plant is a large, continuous process for which conventional analog control would perform as well as DDC on most of the control loops.

Phillips' results should emphasize the fact that computers do not automatically generate economic returns.

1-5 SUPERVISORY COMPUTER CONTROL

The basic objective of a process operation is to optimize the financial return on investment. The economic return on an operation depends upon a number of factors, one of the significant ones being the day-to-day operating strategy. It is frequently not obvious to the operating personnel what the optimum strategy should be. A plant is a complex, interacting entity, and the optimum operating strategy can only be ascertained after considering the combined effect of many different options.

The obvious approach is to use the digital computer to perform just such an analysis. Typical input information needed might include

1. Cost of raw materials and utilities.
2. Value of products.
3. Composition of raw materials and products.
4. Current values of variables within the process.
5. Constraints on the operation (e.g., safety limitations).
6. Specifications on products.

A process model is needed to relate all of these various factors to the economic return on the operation. The optimum operating strategy is that which optimizes this return.

Although the computer determines the optimum operating strategy, the analog control system still implements the decisions. Thus in many cases the control computer simply provides the set points for the analog control loops as illustrated in Fig. 1-5. The computer system does not replace any analog hardware. The backup problem

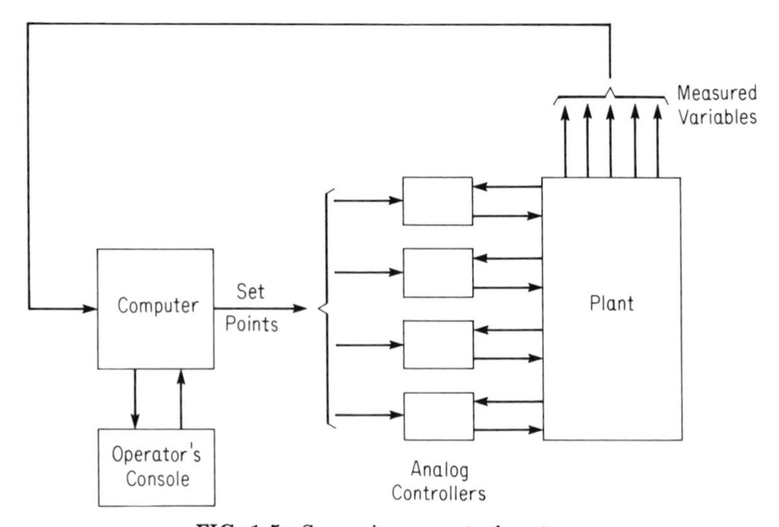

FIG. 1-5. Supervisory control system.

is not as critical, for in case of computer failure the set points simply remain at their last settings.

The concept of supervisory control actually dates back to the development of autopilots for aircraft. Early development was conducted by the Bunker-Ramo Corporation and by Hughes Aircraft, the latter holding a patent, called the Exner patent (13), covering the supervisory concept (specifically, using a digital computer to provide the set point for an analog controller). Bunker-Ramo entered the process computer field, and this company and its successors pioneered much of the early development.

The justification of a supervisory control computer has been a very difficult undertaking and may very well remain so. There are simply too many unknowns prior to initiation of the project. Particularly in the early ventures, it can jokingly be said that the decisions were made in much the same manner as one buys an automobile. That is, they first decided that they wanted a computer, and then looked for the justification. Quite frequently the project was charged against a research budget, thus to some extent avoiding the question of economic justification.

While this justification is not easy beforehand, it turns out to be almost as difficult afterwards. It is simply not obvious which improvements in process operation can be charged to the computer and which cannot. Depending upon who you talk to, the same installation can be a fabulous success or an outright failure.

In most cases, the literature is of little help in justifying a venture. The various companies simply do not publish whatever economic results are obtained from a venture. Probably the one factor that best judges a project is the answer to the question, "Have they ordered another computer?" The answer to this question has in many cases been yes, the rate of placing orders being limited primarily by the personnel available to install them.

The economics of supervisory systems are based on the prospect of the system producing sufficient improvement in process operation to justify the financial investment in the computer control system. This is seldom easy to verify beforehand, but processes that are likely to give such a return generally exhibit one or more of the following characteristics (14,15):

1. *Plants with large throughputs.* Examples include paper machines, power plants, pipe stills, etc. With the very high throughput, only a small percentage improvement is spread over a high volume to generate a large return.
2. *Very complex plants.* In such plants it is very difficult for the operating personnel to assimilate all the various factors to

arrive at the optimum operating objectives. Although the control computer is ideally suited for this, an accurate mathematical model is a must.

3. *Plants subject to frequent disturbances.* Although the first-level control systems can compensate for many of these, the operating objectives must be modified for others, a task for which the frequent attention of the computer will be superior to the operator's performance. These disturbances may be either physical (e.g., changes in feed stock or ambient conditions) or economic (e.g., changes in a raw material or product value).

While this list does not encompass all successful installations, it will include by far the majority.

The main obstacle to the installation of supervisory systems is that mathematical models of plants are seldom available beforehand. Thus the project must justify the expenditure of funds for this effort, which is by no means minimal. This effort typically requires the work of several engineers for periods of a year or more, plus plant tests and additional laboratory data. In many cases the computer is installed early in order to promote this work. It can easily amount to 25 percent or more of the total project cost.

1-6 THE HIERARCHY CONCEPT

The preceding two sections presented two distinct approaches toward the application of computer control to process units. In reality the final system is usually a hybrid between the two, containing parts of both. It seems that most applications have been supervisory in nature, but still incorporating a few of the really attractive DDC loops.

The above discussion presented supervisory control as a digital computer providing the set points for analog control loops. However, the same principle applies to the case in which the supervisory digital computer supplies the set points to the direct digital control computer. In fact, direct digital control becomes even more attractive in these cases because of these reasons:

1. Few if any additional analog inputs are required. Thus this cost appears only once in the total configuration.
2. The communication between the two digital computers is considerably superior to the communication between the digital and analog systems.
3. The backup problem in DDC can be solved by allowing the supervisory computer to assume control of critical loops in case of failure of the DDC computer.

4. The Exner patent apparently does not cover this application. This concept has certainly been promoted by some of the new features of third generation computers, such as multiple ports to memory.

The extension of these concepts to higher and higher levels gives the "hierarchy concept" or the "automated company" (16,17), as illustrated in Fig. 1-6. The lowest level is occupied by the DDC computer, being responsible for the control of a single plant. On the next step is the supervisory computer, which may be responsible for several individual plants.

The next step up the ladder is to a computer responsible for coordinating an entire complex of several plants. In such a complex,

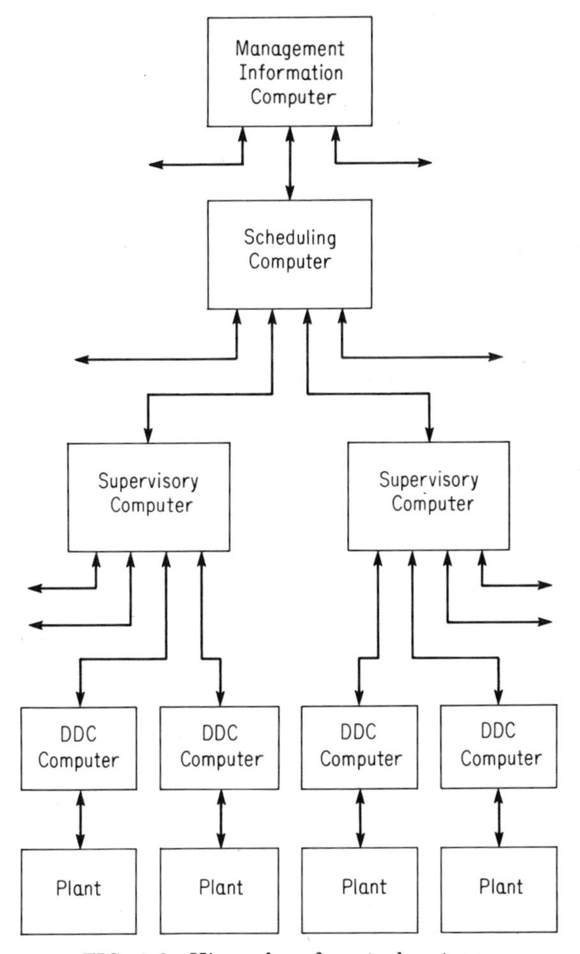

FIG. 1-6. Hierarchy of control systems.

the shipping of materials from one plant to another means that the operations in one are highly dependent upon the operations in another. It also follows that what is best for an individual plant in the complex may not be best for the complex as a whole. As the company is only interested in the total return on investment from all the plants, there may be enough incentive to install a computer to ascertain what operating goals in each plant provide the maximum return from the whole complex. These goals are forwarded to the supervisory computer, which in turn calculates the set points for the DDC computer.

At the top of the ladder is the corporate level control computer, which should logically be a part of the management information system. This computer makes available to management an up-to-date status report on the operation of the entire company. On the basis of current market information, it could have the ability to make some policy decisions, which are then communicated to the computers on the lower levels. It could also receive other policy decisions from management and likewise pass them along. In fact, one of the incentives to install such a system is that decisions at the corporate level could be disseminated to the individual plants in a matter of hours instead of the customary weeks or months. In this way large companies could be as responsive as small ones to market conditions.

LITERATURE CITED

1. Murrill, P. W., *Automatic Control of Processes*, Intext Educational Publishers, Scranton, Pa., 1967.
2. Shinskey, F. G., *Process Control Systems*, McGraw-Hill, New York, 1967.
3. Ziegler, J. G., and N. B. Nichols, "Optimum Settings for Automatic Controllers," *Trans. ASME*, November 1942, p. 759.
4. Wills, D. M., "A Guide to Controller Tuning," *Control Engineering*, August 1962, p. 93.
5. Smith, C. L., and P. W. Murrill, "Controllers—Set Them Right," *Hydrocarbon Processing*, February 1966, p. 50.
6. Lopez, A. M., et al., "Tuning Controllers with Error-Integral Criteria," *Instrumentation Technology*, November 1967, p. 57.
7. Wherry, T. H., and J. R. Parsons, "Guide to Profitable Computer Control," *Hydrocarbon Processing*, April 1967, pp. 179–182.
8. Bernard, J. W., and J. F. Cashen, "Direct Digital Control," *Instruments and Control Systems*, September 1965, p. 151.
9. Bruce, R. G., and R. J. Fanning, "How Direct Digital Control Works," *Hydrocarbon Processing*, December 1964, p. 83.
10. Guisti, A. L., R. E. Otto, and T. J. Williams, "Direct Digital Computer Control," *Control Engineering*, June 1962, p. 104.

11. Rosenbrock, H. H., et al., "Computers and Direct Digital Control," *Control*, March 1965, p. 125.
12. Parsons, J. R., M. W. Ogleby, and D. L. Smith, "Performance of a Direct Digital Control System," presented at the 25th Annual ISA Conference, Philadelphia, Oct. 26–29, 1970.
13. Aronson, A., "Hughes-Owned Patent Looms Large in Process Industries," *Control Engineering*, March 1965, p. 21.
14. Stout, T. M., "Computer Control Economics," *Control Engineering*, September 1966, p. 87.
15. ——, "Estimating Plant Profits for Process Computer Control," *Instrumentation Technology*, June 1969, p. 56.
16. Hodge, B., "Company Control Via Computer," *Chemical Engineering*, June 7, 1965, p. 177.
17. Gaines, N. W., et al., "Union Carbide Integrates Multi-Computer Process Control," *Instrumentation Technology*, March 1967, p. 49.

chapter **2**

The Computer Control System

The objective of this chapter is to briefly discuss the hardware (both computer and computer/process interface) and software generally found in a process computer configuration. Since other texts are available, we shall not give a very detailed discussion of subjects such as how a digital computer works. Furthermore, computer hardware has historically changed very rapidly, so any discussion is likely to become obsolete very quickly. In this chapter our main objective is to try to show the relationship of various hardware and software features to the capability of the computing system to perform in a process control environment.

Discussion of specific systems is intentionally avoided.

2-1 NUMBER SYSTEMS

The smallest storage unit in a digital computer is called a *bit*, a contraction of "*bi*nary dig*it*." It can assume only two states—on or off—and thus can represent only the numbers zero and one. The base two or binary number system is most conveniently and efficiently used in such computers, which are frequently referred to as *binary machines.*

While the machine may conveniently work with binary numbers, programmers do not find this representation especially convenient. A casual examination of the first column of Table 2-1 should reveal the reason: too many ones and zeros leads to confusion. Unfortunately, conversion to the common decimal or base 10 number system is not especially easy. Instead, conversion to the octal (base 8) or hexadeci-

TABLE 2-1
Number Systems

Binary (base 2)	Octal (base 8)	Decimal (base 10)	Hexadecimal (base 16)
0	0	0	0
1	1	1	1
10	2	2	2
11	3	3	3
100	4	4	4
101	5	5	5
110	6	6	6
111	7	7	7
1000	10	8	8
1001	11	9	9
1010	12	10	A
1011	13	11	B
1100	14	12	C
1101	15	13	D
1110	16	14	E
1111	17	15	F
10000	20	16	10

mal (base 16) system is quite direct. For example, to convert from binary to octal, simply group the binary digits in groups of three from the right, and convert each group to octal. The binary number 100110111010 is converted as follows:

100 110 111 010
4 6 7 2

Similarly, it is converted to hexadecimal as follows:

1001 1011 1010
9 B A

Conversion from octal or hexadecimal to binary is equally as easy. For the beginner, Table 2-1 is a useful aide, but it becomes unnecessary with a little practice.

Another characteristic that should be noted about the binary number system is the largest decimal number that can be represented by a given number of bits, which is given in Table 2-2 for up to sixteen bits. The first four entries can be verified from Table 2-1. The other entries can be computed as follows:

$$\text{Largest decimal number} = 2^n - 1$$

TABLE 2-2
Number of States per Number of Bits

Number of Bits	Largest Decimal Number	Number of States
1	1	2
2	3	4
3	7	8
4	15	16
5	31	32
6	63	64
7	127	128
8	255	256
9	511	512
10	1,023	1,024
11	2,047	2,048
12	4,095	4,096
13	8,191	8,192
14	16,383	16,384
15	32,767	32,768
16	65,535	65,536

where n is the number of bits. For example, computers that store data as one entry per sixteen bits are common. Reserving one bit for the sign, the largest number that can be stored in the remaining fifteen bits is 32,767. Another way of looking at this is to say that the maximum resolution of this data is one part in 32,767, or 0.003 percent.

In other applications, the number of states that can be represented by n binary bits is of importance, which is also given in Table 2-2. This is simply one more than the largest decimal number.

In the second generation computing machines (IBM 7094 and similar series), six bits were sufficient to represent the character set (letters of the alphabet, the ten digits, and special symbols such as the decimal point, comma, parentheses, etc.). Two octal digits could represent the six bits, and the use of the octal number system was common. With the introduction of the next generation of computers (IBM 360 and similar series), the character set was expanded, requiring eight bits for representation. The term "byte" arose to refer to a group of eight bits, and such computers were often referred to as byte-oriented machines. As two hexadecimal digits are required to represent the eight bits in a byte, this number system began to be used in place of the octal system. Not all manufacturers adopted the expanded character set, so the octal system still enjoys some use.

Actually, the expanded character set is not necessary for most process control systems, but it is convenient for compatibility with the larger data-processing machines.

2-2 CENTRAL PROCESSING UNIT

The central processing unit, often designated CPU for short, is the heart of the computer, as illustrated in Fig. 2-1. Among its

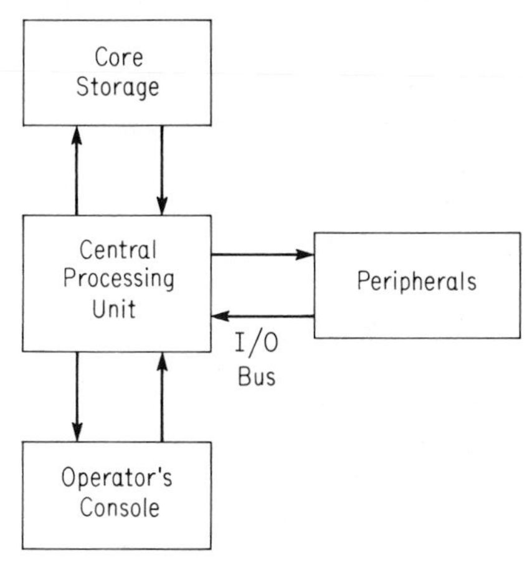

FIG. 2-1. Schematic representation of a computer.

primary functions are the following:

1. Keeps track of the current location in the sequence of instructions via the *instruction address register*, which generally contains the address of the next instruction to be executed.
2. Retrieves instructions from core storage, decodes them, and executes them. The CPU contains hard-wired logic to perform a certain number of operations, which comprise the instruction set for the computer. These instructions might entail storage or retrieval of data from core storage, arithmetic operations, logic operations, or shift operations.
3. In simpler machines the CPU is responsible for the transfer of data between core storage and the peripheral units. In more sophisticated machines the CPU only directs these operations, a point we shall examine more closely in a subsequent section.

The word length of the computer generally corresponds to the number of bits which the processor stores in or retrieves from core storage in one read/write operation. Word lengths vary from machine to machine, with 8-bit, 12-bit, 16-bit, and 24-bit word lengths commonly used in process control computers. Of these, the 16-bit word length is most common.

The address of a word designates its location in core storage. Given the address, the CPU can retrieve its contents from core storage. However, the contents of the word generally give no clue as to the address from which it came.

The cycle time of the machine is the time required for the CPU to read one word from core storage and restore the contents. The cycle time is basically determined by the size of the ferrite rings used in the core storage on current computers. The smaller these rings the faster the machine. But as the rings become smaller, the energy required to energize or de-energize them becomes smaller, and thus faster core is more subject to noise-induced errors. Cycle times on current machines range from slightly less than one microsecond (μsec) to about 4 μsec.

As we shall see, the cycle time is not the sole determinant of how fast the computer will execute a given set of code. For example, not all instructions can be executed in one cycle time. Furthermore, the instruction sets differ considerably from one machine to the next. Therefore a task that one machine could accomplish by executing one or two instructions might require four or five on another machine. Even though the second machine might have a shorter cycle time, it may not perform the desired operation as fast as the first machine.

To assist in performing various operations, the CPU has a number of registers, one of which, the instruction address register, has been mentioned already. Earlier machines had separate registers for different purposes, such as an accumulator to store the results of arithmetic operations, index registers for modifying addresses, and other registers for various purposes. Current machines tend to have general-purpose registers which can be used for practically any purpose with few restrictions. In this way, the registers are of more general utility and enable the programmer to prepare a more efficient program. All other things being equal, a computer with more registers will generally perform a given task faster than a machine with fewer registers.

Preferably, the registers are implemented as flip-flops in the CPU itself. An alternative is to reserve a few storage locations in the lower part of core storage for use as registers. This leads to a less expensive CPU but also to slower execution speeds. When a register is part of core, one memory cycle time is required to retrieve its contents, whereas considerably less time (on the order of 200 nanosec (0.2 μsec) or less) is required when the registers are part of the CPU.

A feature now enjoying considerable popularity is the read-only memory (ROM), a medium in which information is stored in permanent (nonerasable) form. This type of storage offers three advantages over read/write core.

1. Faster by a factor of about 10.
2. Less expensive.
3. Stored information is permanently protected from erasure by a "run-away" program.

Current practice is for the ROM to be prepared at the factory with field modification virtually impossible, but field-programmable ROM's are expected.

As an example of an application of an ROM, a commonly used routine such as the square root could be implemented in ROM to take advantage of the increased speed of execution. In other applications, special mathematical routines such as the fast Fourier transform could be implemented via ROM.

Microprogramming is another feature that increases the flexibility and decreases the costs of the central processor, making it quite popular for use in small computers. In this approach, a microprogram is prepared giving the elementary sequence of steps required to perform the same instruction that otherwise would have been implemented as a hard-wired instruction. In this approach, microprograms could be prepared to enable one machine to execute the instructions of another machine (i.e., to *emulate* the second machine). Use of an ROM in which to code these instructions is certainly advantageous.

2-3 RELATIONSHIP OF WORD LENGTH TO PERFORMANCE

When selecting a computer, the user can choose between various machines with different word lengths. For process control, the 12-, 16-, 18-, or 24-bit word lengths are all frequently used. The word length has a definite impact on the performance of the computer, and thus becomes an important factor in machine selection.

As either a data entry or an instruction can be stored in a word of memory, consideration must be given to both. We shall first consider data storage, then the instructions.

Process data generally enters the computing system in integer or fixed-point format. For example, suppose the input is a voltage signal in the 0 to 5 volt d-c range. If we use an 11-bit A/D converter, an input of 0 volts would correspond to all bits being set at zero; an input of 5 volts would correspond to all bits being set to 1, giving the binary representation of the decimal number 2047 (refer to Table 2-2). Since the resolution of this arrangement is 1 part in 2047 or slightly better than 0.05 percent, this is entirely adequate for most process transducers, whose accuracy is usually about 0.1 percent. Adding a bit for the sign gives a total of 12, and therefore a 12-bit word length would be adequate for storing most process data in integer format. Use of a longer word length would be wasteful.

When working with process data, it is generally more convenient to first convert it to engineering units. The integer or fixed-point representation is not especially convenient for this purpose, the real (floating-point or exponential) format being much more attractive. In this approach, a certain number of bits are reserved to represent the characteristic (including sign) and a certain number of bits are reserved to represent an integer exponent (including sign). The minimum workable combination is to reserve about 18 bits for the characteristic (giving from four to five decimal digits of precision) and about 6 bits for the exponent (which is sufficient to represent numbers between approximately 10^{-9} and 10^{+9}). This requires a total of 24 bits.

Although four digits is generally sufficient to represent the raw process data, this relatively low precision coupled with the round-off characteristics of binary machines often leads to numerical problems even in relatively simple mathematical procedures. Using a total of 32 bits, giving seven or eight digits of precision, to represent a real number circumvents these problems in most process control applications.

Insofar as process control applications are concerned, the following general statements apply to the selection of the word length in light of the data storage aspect.

12-bit word. Since two or three words would be required to represent a floating-point number, virtually all data must be stored in integer form. In fact, floating-point operations should be avoided. Therefore, machines in this category could be considered only for those applications in which little or no floating-point operations are expected.

16- or 18-bit word. In these machines, the use of two words to store a floating-point number makes their use a bit inconvenient but yet quite feasible. Storage of data in integer format reduces the words of core storage required by a factor of two. Manipulations of floating-point data will also be inherently slower because two memory cycles are required to retrieve a floating-point number from core storage as compared to one cycle to retrieve an integer number.

24-bit word. In these machines there is no penalty for storing data in floating-point format. However, the relatively low precision of the floating-point number may require the use of double precision in some operations.

Of course, the cost to performance ratio is really the number of importance. Currently (1971), core storage costs about one dollar per byte (8 bits). Naturally, the 24-bit word length is the more expensive.

Virtually all process control computers in use today employ some variation of the single-address instruction format. As illustrated in Fig. 2-2, the instruction is divided into three fields, the operation

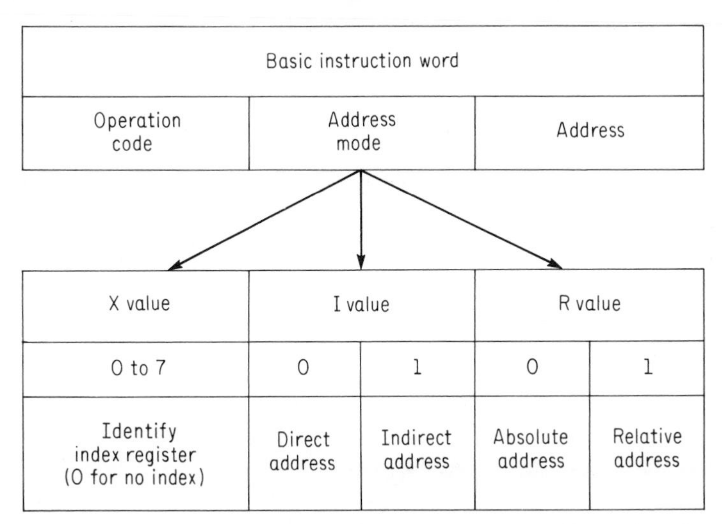

FIG. 2-2. Format of a single-address instruction. (Reproduced by permission from Ref. 1.)

code, the address mode, and the address itself. The purpose of these fields are as follows:

Operation code. This field specifies the operation to be performed.

Address. This field contains the address utilized in executing the instruction.

Address mode. This field designates what modifications are to be made in the address contained in the address field before the instruction is executed.

Disregarding the address modifications for the moment, consider the following examples of instructions:

Left-Shift or *Right-Shift.* This instruction causes the contents of the accumulator to be shifted to the left by one bit or to the right by one bit. The address field is not used.

Store-Word. This instruction stores the contents of the accumulator into the word whose address is in the address field. The converse of this operation is "load word."

Unconditional Transfer. After execution of this instruction, the next instruction executed is the one whose address is in the address field of the transfer instruction. Execution of the transfer instruction simply requires placing the contents of the address field into the instruction address register.

Load Immediate. Some instructions treat the contents of the address field as if it were data. For example, the "load immediate" instruction transfers the contents of the address field into the accumulator.

This last instruction illustrates an example of the effect of the instruction set on the machine's performance. On machines with an abbreviated instruction set not containing the "load immediate" instruction, a word of core storage must be reserved for the data and a "load word" instruction used instead. This "wastes" a word of core storage.

In direct addressing, the address field contains the actual address of the information to be accessed. In process control computers, three common approaches to modifying this address are used:

1. *Relative Addressing.* The contents of the address field are added to the contents of the program-location register to obtain the address to be used. In computers without this feature, a program is written (or compiled) to be executed from a predetermined location in core storage. Incorporating this feature permits a program to be loaded into any position in core storage and executed, a feature called *dynamic allocation of core storage.* As we shall see in a later section, this can be done only with the aid of a mass-storage device such as a disk or drum. Therefore, this feature is of little value on all-core machines.
2. *Indirect Addressing.* In its simplest form, the address field contains the address of a word in core storage that contains the address to be used in executing the instruction. This is known as *single-level* indirect addressing. This procedure can be nested to give *multilevel* indirect addressing. An extra memory cycle is required for each level of indirect addressing.
3. *Indexed Addressing.* The contents of an index register are added to the contents of the address field to obtain the address to be used in executing the instruction. If the index register is implemented as a word in core storage, a memory

cycle is required to retrieve its contents. Implementing the index register as flip-flops in the CPU saves this time.

All of these types of address modifications may be used simultaneously.

To illustrate the effect of word length on the computer's performance, suppose we are considering a 16-bit machine with three index registers and the capability to perform relative and indirect addressing. This means that the address mode field must contain four bits—two bits to designate the index registers, one for relative addressing, and one for indirect addressing. This leaves twelve bits for the other two fields.

Furthermore, suppose four bits are reserved for the operation code. Table 2-2 indicates that four bits can designate only 16 different instructions, a rather paltry number. However, ingenious schemes have been devised to circumvent this problem. For example, all instructions not utilizing the address field are given the same operation code. Then the contents of the address field are used to specify the specific operation to be performed.

Reserving four bits for the operation code and four bits for the address mode leaves eight bits for the address field. Table 2-2 indicates that eight bits would be sufficient to direct address only 256 words of core storage. This fact indicates that indirect addressing must be used extensively on these machines, thereby reducing the effective speed with which they can execute a program.

On 24-bit machines the address field is sufficient to direct-access about 16K (K = 1,024) words of core storage. Thus indirect addressing is used less frequently. On 32-bit machines, the address field is generally sufficient to direct-access all of core storage.

On machines with word lengths shorter than 16 bits, double-word instructions must frequently be used, thereby offsetting the advantages of using the shorter word.

As the final point in this section, it should be noted that the word length essentially fixes the maximum core storage available on a 16-bit machine. As the maximum address that can be represented by 16 bits is 65,535, the maximum core available on most 16-bit machines is 64K.

2-4 CPU OPTIONS

In this section, we shall define a CPU option as any feature of the CPU that is optional on some (not all) computers that are frequently considered for process control. That is, some of our "options" are standard features on some computers.

Hardware Multiply/Divide (Also Called Fixed-Point Arithmetic)

Virtually all CPU's have an instruction to add the contents of a memory location to the contents of the accumulator (i.e., a fixed-point add instruction). While multiplication of two fixed-point numbers can be accomplished by successive additions and shifting operations, this entails two penalties:

1. Execution speed is reduced due to the large number of operations required.
2. The instructions required in this procedure must be stored at least once (usually as a subroutine) in core storage.

Division can be accomplished in a similar fashion, and the software routines for this purpose are commonly referred to as fixed-point software.

An alternative procedure is to implement hardware to perform fixed-point multiplications and divisions. This eliminates the need for the software and also increases execution speeds significantly, the order of magnitude being as follows:

	Hardware	Software
Multiply	10 μsec	200 μsec
Divide	20 μsec	500 μsec

As the cost is also reasonably low (about $2,000 in 1971 prices), this feature is found in most process control computers. However, in computers used for other purposes (e.g., in communications networks), this feature is not so important.

Hardware Floating-Point Arithmetic

In the minimal configuration, few CPU's have the capability to perform any floating-point operation. Just as in the case of fixed-point multiply/divide, either software routines may be used or additional hardware can be purchased. In either case, the functions that must be supplied include addition, subtraction, multiplication, division, and other floating-point manipulations. Orders-of-magnitude comparison of execution speeds of hardware vs. software are as follows:

	Hardware	Software
Add and Subtract	15 μsec	400 μsec
Multiply	20 μsec	400 μsec
Divide	30 μsec	1000 μsec

This feature is not commonly found on process computers because 1) the price is substantial (about $20,000 or more in 1971 figures), and 2) floating-point operations can be avoided to a large extent on process control computers.

Storage Protect

In process control computers, it is frequently desirable to protect a certain segment of the programs from being accidentally written over by a runaway program outside this segment of programs. One approach to implement this is by including a protect bit with each word of core storage. In this way a protected location of core storage can be written into only by an instruction whose protect bit is on. This feature in some form is found on most process control computers.

Because of the expense of adding a bit to each memory location, some manufacturers have adopted the paging concept for storage protect. In this approach, a single protect bit is provided for a segment of core storage generally consisting of about 256 or 512 words, otherwise known as a page.

Parity

In order to provide some error-detection and correction capability, a parity bit can be added to each word of core storage and to words of information transferred between peripheral devices. To illustrate the functioning of parity, suppose the parity bit is set "on" when the number of "on" bits in the word is odd. If an even number of bits are "on," the parity bit is set "off." Then including the parity bit, the number of bits that are "on" should always be even. If an error is made involving any one bit, the number of "on" bits would be odd, indicating an error. If two errors are made they would not be detected, but the probability of this happening is extremely remote.

Several manufacturers, contending that their core storage is so reliable that parity checking is not needed, do not even offer it as an option. However, peripherals are not so reliable, and data transferred to and from peripherals should always be accompanied with a parity bit.

Real-Time Clock

Virtually all process control computers require a real-time clock in order to coordinate the computer's operation with the real world's time schedule.

Power Fail-Safe

In the event of loss of power to the computer, this option provides the capability of executing a set number of instructions before the machine becomes inoperable. These instructions may generally be used for whatever the specific application requires.

Automatic Restart

With loss and resumption of power, the contents of core storage are not altered. However, the contents of the working registers implemented as flip-flops in the CPU are lost. But if some of the instructions available from the power fail-safe option are used to store the contents of the working registers, program execution can proceed when power is resumed. The function of the automatic restart option is to reload the working registers with their contents at the time of loss of power and resume program execution.

Watchdog Timer or Operations Monitor

If for any reason a program became "hung up" in a never-ending loop, the process control computer would effectively cease to perform all needed functions. To provide protection against this, the watchdog timer must be reset within a certain allotted time period (e.g., 15 sec) by whatever program or programs are being executed. Failure to do this serves as an indication of a problem somewhere in the software.

2-5 I/O STRUCTURE

As indicated previously, input/output (I/O) operations in earlier computers were accomplished via the CPU. In this way the CPU was committed to the I/O operation while it was in progress, and therefore was not available for other functions.

The I/O performance was improved by adding an I/O processor which operated independently but yet through the CPU on a cycle-stealing basis. That is, the CPU instructed the I/O processor as to what operations were needed, and these were performed by "stealing" memory cycles from the CPU as the peripheral device could receive or transmit information. This frees the CPU so that the remaining memory cycles can be used for computational purposes.

By using a multiple port to memory or direct memory access channel as illustrated in Fig. 2-3, the CPU is completely free of the

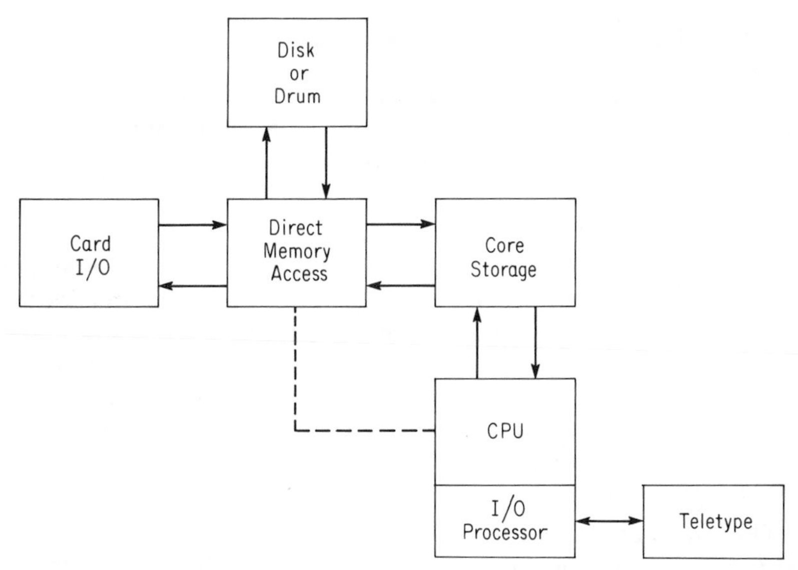

FIG. 2-3. Direct memory access channel.

major I/O functions. The direct memory access channel essentially consists of a satellite CPU whose functions are basically limited to I/O operations. When high data-transfer rates are expected, this approach is extremely attractive.

The use of multiple ports to memory can produce a variety of computer configurations, even involving multiple processors as illustrated in Fig. 2-4. Each CPU has its own private memory in addi-

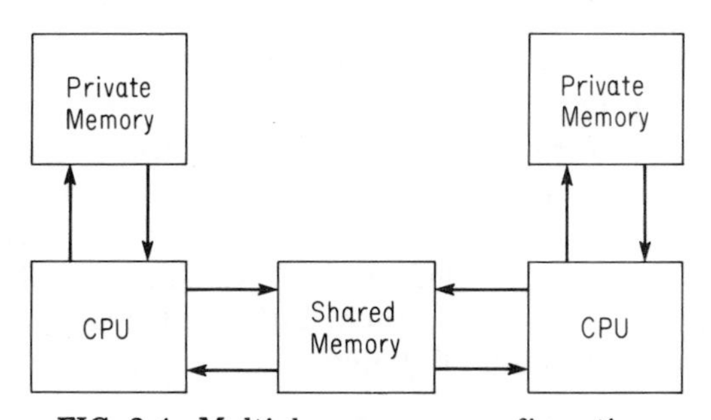

FIG. 2-4. Multiple processor configuration.

tion to the shared memory, which enables the two processors to communicate with each other quite readily. Peripherals with or without a direct memory access channel can be added to each CPU.

2-6 PERIPHERAL DEVICES

In this section we will be concerned only with the classical data-processing peripherals—teletype, paper tape, and similar devices. Process-oriented I/O devices are discussed in a later section.

Teletype

Virtually all computers have a teletype or typer in the computer room for communications with the computer operator. In addition, many process control computers have additional teletypes or typers out in the field for operator communications. These devices are rather low speed (10 to 15 characters per second), but their low cost makes them quite attractive where the output volume is low.

CRT Display Units

The low-speed output from the teletype detracts from its utility for operator communications. When a hard copy is not necessary, the cathode-ray tube (CRT) display units can accept a rather high data rate, and therefore are becoming quite popular for operator communications. One approach is to display information to the operator via the CRT, obtaining a hard copy of the desirable information via the teletype or line printer in the computer room. The alphameric CRT's are reasonably cost-competitive with the teletypes. Vector-drawing CRT's are considerably more expensive and therefore used more sparingly.

Paper Tape Read/Punch

While a slow-speed (10 characters/sec) paper tape read/punch can be added to a teletype for a nominal expense, the input/output speeds are too slow for all but a few applications. A high-speed paper tape unit (200 characters/sec reader; 100 characters/sec punch) has sufficient speed for normal program preparation, program debugging, and system maintenance. This unit is substantially less expensive than an equivalent card read/punch, but is not nearly as convenient for program preparation and debugging.

Card Read/Punch

While the high-speed paper tape unit was rather standard on early process control computers, the card read/punch has replaced it on practically all systems on which a significant program development effort is anticipated. Typical speeds for card read/punch units on process control computers are 200 card/min reading, 80 card/min punching. A card I/O unit generally costs at least twice that of a comparable paper tape unit.

Line Printer

The volume of printed output from a process control computer is seldom sufficient to justify the cost of a line printer. But for systems on which a large program development effort is expected, consideration should be given to renting a line printer during the initial programming stages when the volume of output is high.

Drum

A drum is a mass-storage device on which information is stored on the magnetized surface of a rotating drum. This surface is divided into tracks with a read/write head over each track. The rotating speed of the drum is such that one revolution is made every 33 millisec. If the item of information to be read from the drum has just passed under the read/write head, the computer must wait 33 millisec for the drum to make a complete revolution. This is the worst possible case, and is known as the *maximum access time*. On the average, the computer would have to wait for the drum to make one half revolution or 17 millisec, which is referred to as the *average access time*. The read/write circuitry is fast enough so that words can be read from or written onto the drum sequentially as it rotates.

The advantages and disadvantages of a drum relative to a disk are discussed in the next section.

Disk

A disk is similar to a drum except for two aspects. First, the magnetic coating is on the surface of a flat, circular plate which rotates at about the same speed as the drum. Second, most disks are equipped with a single head that can be moved from track to track to obtain the desired information. The average access time of the disk is essentially dictated by the speed of the positioner. Disks with very slow mechanical positioners have an average access time of about 500 millisec and a maximum access time of about twice this. Disks with the very best positioners have average access times of around 100 millisec. Recently disks have appeared with a read/write head per track. With an average access time of about 17 millisec, these disks are virtually equivalent to drums, and are often referred to as *drisks*.

The relative advantages and disadvantages of a movable head disk over a drum or drisk are

1. Since the read/write heads are quite expensive, the disk is

generally less expensive than a drum of the same storage capacity.
2. As evidenced by the average access times listed previously, the drum is faster.
3. Mechanical components have historically been the least reliable portion of a computer system. Thus by eliminating the mechanical positioner the drum is generally more reliable.
4. Many disks permit disk surfaces to be interchanged, which permits a copy of the information on the disk to be stored off-line as a backup. This is not possible with drums or drisks.

Maximum storage capacity is generally not a factor, since very large disks and very large drums are available. For process control, a minimum of a million words is generally required.

Magnetic Tapes

Due to the comparatively long access time of the magnetic tape, these units are rarely found on process control computers.

2-7 TYPICAL CONFIGURATIONS

Process control computers come in a wide variety of configurations, depending heavily on the application. In this section we shall give typical configurations of three classes of computers. For each of these we shall give an approximate cost breakdown based on 1971 prices.

Minicomputer

Usually installed as a dedicated computer to perform a relatively simple task, the configuration, as illustrated in Fig. 2-5, is practically

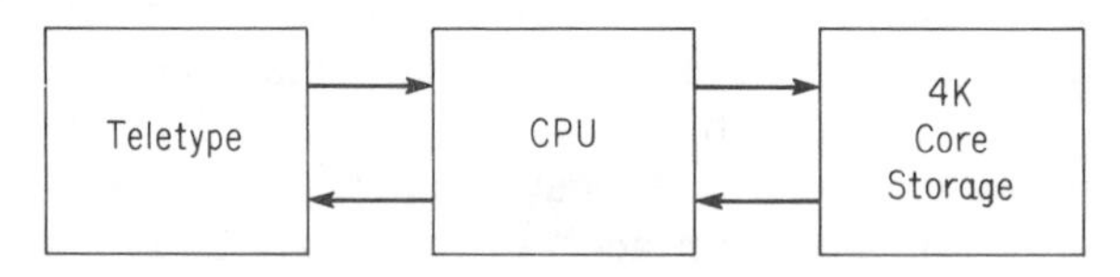

FIG. 2-5. Configuration of a minicomputer.

the absolute minimum. These machines are generally programmed in assembly language on a once-for-all basis. For this to be practical, the task the computer is to perform must be well-defined beforehand.

Most writers tend to define the minicomputer in terms of its cost (2). A typical definition of a minicomputer is one costing less than $25,000, again in 1971 prices. The configuration in Fig. 2-5 could be purchased in 1971 for less than $15,000 even with a 16 bit word length.

Direct Digital Control

Figure 2-6 illustrates a typical configuration of a computer used for direct digital control. Because fast response is generally the basic

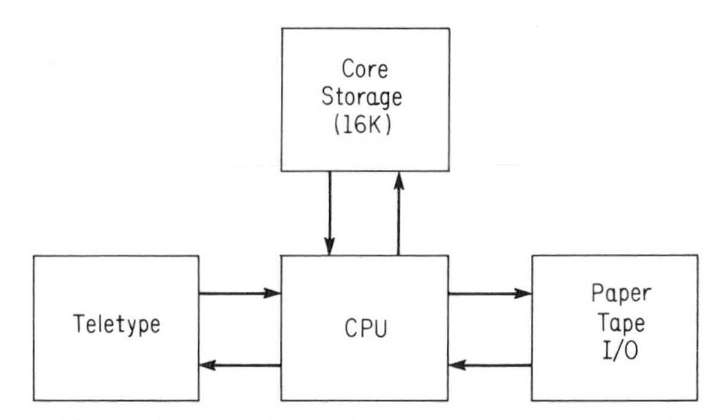

FIG. 2-6. Typical configuration of a direct digital control computer.

requirement of a DDC system, an all-core (no disk or drum) machine is illustrated. Programming would generally be in assembly language, although several standard DDC packages are available. Since relatively little programming effort is anticipated after the system once becomes operational, a paper tape I/O is frequently used on these machines.

Based on 1971 prices, the cost of the configuration in Fig. 2-6 is approximately as follows:

CPU (16 bit) with hardware multiply/ divide, storage protect, real-time clock, power fail-safe	$20,000
Core storage	32,000
Teletype	3,000
Paper tape I/O	8,000
	$63,000

A machine of this configuration would probably be adequate for no more than 100 loops with a reasonable complement of feedforward, cascade, and other advanced control strategies.

Supervisory Systems

On the configuration of the supervisory system illustrated in Fig. 2-7, most of the programming could be done in a compiler level

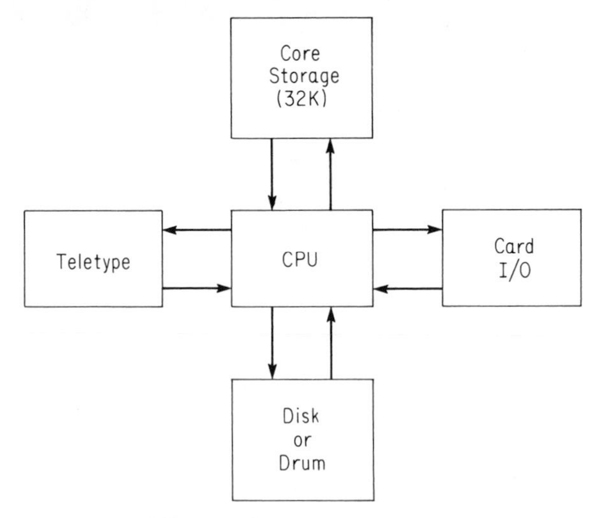

FIG. 2-7. Supervisory system.

language such as Fortran. Program development and compiling could be done on-line. The operating system would transfer programs from the disk or drum to core storage for execution. Since program development and subsequent system improvement is likely to continue for some time, card I/O is preferred.

Based on 1971 prices, the cost of the configuration in Fig. 2-7 is approximately as follows:

CPU (16 bit) with hardware multiply/ divide, storage protect, real time clock, power fail-safe	$50,000
Core storage	64,000
Card I/O	20,000
Teletype	3,000
Disk	35,000
	$172,000

Adding a line printer would cost another $30,000, the main reason that it is usually omitted.

Total Project Cost

Although the computer system is generally a significant element in the total project cost, many other factors must be considered to arrive at a total project cost, which includes the computer system and its auxiliary equipment, programming, process analysis, installation, and training. Table 2-3 gives the cost breakdown on eight selected

TABLE 2-3

Costs of Selected Process Computer Systems in Survey Plants with Purchased Computers. (Reprinted from "Outlook for Computer Process Control", U.S. Department of Labor Bulletin 158, 1970.)

Type of Application	Total cost	Computer and Auxiliary Equipment[1]		Programing and Systems Analysis[2]		Installation and Additional Instrumentation[3]		Training[4]	
		Amount	Percent of total system cost	Amount	Percent of total system cost	Amount	Percent of total system cost	Amount	Percent of total system cost
Multicomputer system controlling all major processes in large chemical plant	$1,500,000	$1,125,000	75.0	$225,000[5]	15.0	$150,000	10.0	—	—
Complex system for control of an electric generating station	850,000	400,000	47.1	190,000	22.4	250,000	29.4	10,000	1.2
Operator guide control over a major process in a steel plant	810,000	290,000	35.8	300,000	37.0	200,000	24.7	20,000	2.5
Operator guide control of electric generating station	720,000	300,000	41.7	140,000	19.4	275,000	38.2	5,000	0.7
Direct digital control of a chemical process	500,000	275,000	55.0	75,000	15.0	150,000	30.0	—	—
Control over a key portion of a chemical process (early installation)	453,000	258,000	57.0	75,000	16.6	110,000	24.3	10,000	2.2
Control of analytical instruments in chemical plant laboratory	235,290	160,000	68.0	58,820	25.0	16,470	7.0	—	—
Experimental direct digital control system using 2 computers in a chemical plant	222,000	157,000	70.7	50,000	22.5	10,000	4.5	5,000	2.3

[1] Central processor, auxiliary memory, analog/digital signal converters, and input/output equipment such as operator console, typewriters, and tape equipment.

[2] Analysis of process, preparation of process model, programing for process control, and system operation.

[3] New instrumentation needed for process control installation of computer equipment, and instrumentation including site preparation.

[4] Instructing employees in programing, computer technology, maintenance, and system operation.

[5] Includes training.

process computer systems. Note that the cost of the computer and auxiliary equipment varies from 35.8 to 75 percent of the total system cost. As in all aspects of today's economy, the trend in process control systems is that the hardware costs are tending to decline and the people-related costs are tending to rise.

2-8 PROCESS INTERFACE

In order to function properly, the computer must receive certain data from the process and transmit other data to the process. The computer/process-interface, often called the analog front end, must somehow accomplish these functions. The data involved generally fall into one of the following three categories:

1. Continuous or analog data.
2. Discrete data involving only two levels (i.e., on-off type information).
3. Pulse data.

These catagories apply to both input and output data.

Figure 2-8 illustrates the typical arrangement for reading analog

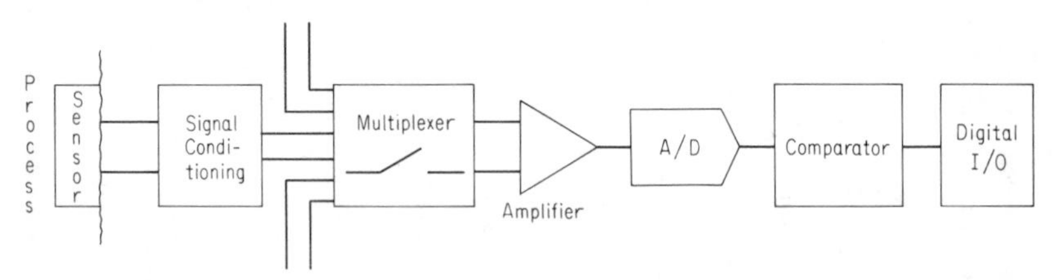

FIG. 2-8. Analog input system.

values from the process. These signals can be classified as follows:

1. Low-level signals, generally considered to be those whose voltage level is less than 100 microvolts (μv), include the outputs of thermocouples, strain gauges, resistance thermometers, and similar transducers.
2. High-level signals, generally considered to be those whose voltage level is greater than 100 μv, emanate from transducers with a built-in amplifier of some type.

Due to the popularity of thermocouples for measuring temperatures, low-level signals are commonly encountered in process control systems. Naturally, these signals are most succeptible to distortion, thus requiring special precautions. The leads generally consist of a twisted, shielded pair. The leads should not be carried in the same tray as a-c power circuits, and in general should not come in close proximity of large electrical motors or generators.

Improper grounding can also be a potential source of distortion of low-level signals. In general, the circuit should be grounded at only one point, preferably at the computer. Figure 2-9 illustrates a

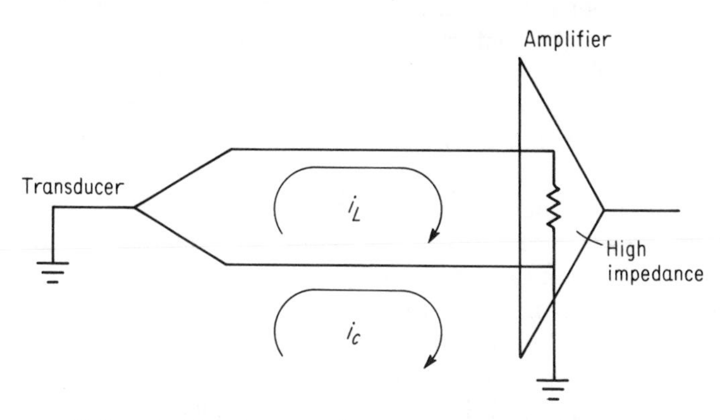

FIG. 2-9. Circuit susceptible to common-mode noise.

circuit grounded at two points, one at the computer (specifically, at the amplifier) and the other at the transducer. The impedance of the amplifier is very large (on the order of 10^6 ohms), so negligible current flows around the loop. However, the two grounds are likely to be a considerable distance apart, and therefore probably at slightly different potentials. Therefore, a current i_c, called the *common-mode current*, flows in one of the leads and not the other (due to the impedance of the amplifier). The voltage drop due to this current causes a bias (which may vary with time) to appear in the reading. This bias is referred to as the *common-mode noise*. As this noise cannot be removed by filtering, steps should be taken to avoid it. The easiest way is to avoid grounding at the transducer.

As rather detailed discussion of good wiring practices are available, they will not be repeated here (3).

The function of each of the elements in Fig. 2.8 is as follows.

Signal Conditioning. This may encompass a variety of elements depending upon the sensor itself. When the output of the transducer is a voltage signal, the signal conditioner generally consists of only an *RC* filter. But if the output of the transducer is other than a voltage signal, the signal conditioner generally transforms it to a voltage signal prior to the multiplexer. For example, if the output of the transducer is a current signal, the signal conditioner generally contains a resistor across which the voltage input to the multiplexer is taken.

Multiplexer. The multiplexer provides the mechanism by which one of several signals is connected to the A/D converter through the amplifier. For high-level signals, solid-state electronics (field-effect transistors) are used in the switching circuits. Sampling rates of 10,000 points per second and higher are readily accomplished. For low-level signals, the distortion of the field-effect transistors cannot be tolerated. Reed- or mercury-wetted relays must be used, resulting in a much slower sampling rate (about 200 points per second). Some systems contain two distinct multiplexers—one for high-level signals and one for low-level signals.

Multiplexers range in size from about 32 input points up to 2,048 input points or more. The sampling sequence on some multiplexers is fixed to a certain sequence, yielding what is called a *sequential scan.* Other multiplexers permit selection of the point to be read, enabling the points to be read in random order. Of course, this latter multiplexer is more expensive.

Both types of multiplexers are found in process control systems. When the computer controls the analog scan, the multiplexer must be capable of reading the points in random order. On other systems, however, control of the scan may reside largely outside of the CPU. Using a sequential scan and a direct memory access channel to store the data in preassigned storage locations relieves the CPU of the burden of supervising the analog scan.

Amplifiers. The function of the amplifier is to scale the process signal either upward or downward so that the resulting range matches that of the A/D converter, typically 15 volts. Some systems utilize a fixed-gain amplifier, in which case voltage-divider circuits often appear in the signal conditioner. In other systems, a programmable-gain amplifier permits the computer to specify which one of several available gains is to be used. This latter alternative provides more flexibility, but the amplifier is more expensive and also requires some output data (i.e., the value of the gain) from the computer.

A/D Converter. Conversion of the signal from analog (continuous) form to digital (discrete) form is accomplished by the A/D converter. The resolution of the A/D converter is related to the number of bits in the digital output by the equation

$$\text{Resolution} = \frac{1}{2^n - 1}$$

where n is the number of bits. For process control, an 11-bit converter is entirely adequate, giving a resolution of about 0.05 percent. For some applications an eight bit converter with a resolution of about 0.4 percent is acceptable.

The time required for the digital output of the A/D converter to reach a constant value after a new input is applied is known as the *settling time.* For solid-state A/D converters, the settling time is 40 μsec or less, which becomes significant only at high data-transfer rates.

Comparator. In order to relieve the CPU of some of its burden, the input data can be compared to high and low limits outside the CPU. This feature is very attractive on systems using the sequential scan coupled with a direct memory access channel to store the input data in preassigned storage locations. The high and low limits are retrieved via the direct memory access channel from preassigned locations in core storage. If either limit is violated, an interrupt is generated, calling for the CPU's attention. Thus the input scan proceeds independently of the CPU until a limit is violated.

Although inputs which can assume only two states could be entered via the route described in the above paragraphs, this condition places an undue burden on the analog input system. Most process control systems permit the states of inputs of this type, known as *discretes*, to be read in groups. Normally each discrete is assigned to a bit in a word. In one cycle time, most computers can read a word containing the status of a number of discretes equal to the word length. The capability to manipulate the bits in a word in order to ascertain which bits are on or off becomes extremely important.

Discretes are commonly used to indicate the status of relays, which may be found in anything from electrical switches to high-pressure alarms. In the convential operator's console, the position of the thumbwheel switches and other devices for data entry is indicated via a bit pattern entered into the computer via discretes.

The output of certain measuring devices such as tachometers or turbine meters is often in the form of pulses. Although the computer can be readily programmed to count pulses for a given length of time, this tends to consume too much of the CPU's time. Instead, external *pulse counters* are generally preferred. In these devices, the CPU loads a register in the pulse counter with the number of pulses to be counted. With the receipt of each pulse, this register is "down-counted" (i.e., one is subtracted, until the register reaches zero, at which time an interrupt to the CPU is generated). To determine the time required for the given number of pulses to occur, the CPU needs only to subtract the time when the pulse counter was initialized from the current time. Thus the CPU has little to do.

The output of data to the process is generally by one of the following three means:

1. Digital-to-analog (D/A) converter, which converts a digital

value (in integer format) to an analog signal. A multiplexer could be used to obtain several outputs from a single D/A converter as illustrated in Fig. 2-10. But with the addition of

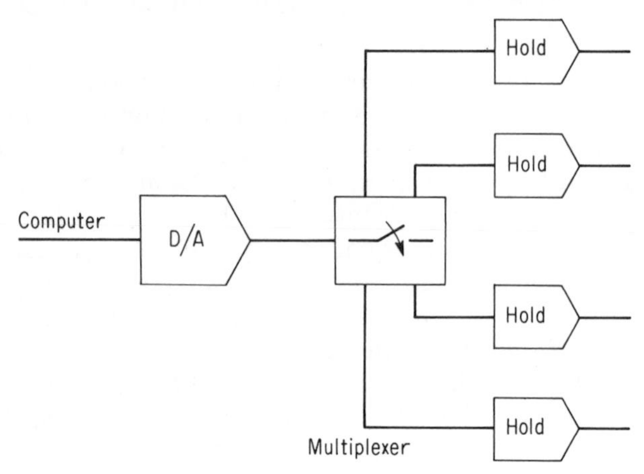

FIG. 2-10. Multiplexing the output of a D/A converter.

the hold circuits to maintain the value between samples, the economics tend to favor individual D/A converters.

2. Pulse generators, which generate the number of pulses specified by the computer. In most systems the pulses are of predetermined amplitude and duration and with a predetermined time between pulses. The outputs of pulse generators are commonly used to drive stepper motors.
3. Contact closures, which can assume only two states—on or off. In addition to simple applications such as turning pumps or lights on or off, a contact can be closed (or opened) for a period of time to obtain a pulse output of variable duration.

To illustrate the use of these devices, consider the output of a quantity such as a valve position or set point for an analog controller. Perhaps the most direct approach is to use a D/A converter as illustrated in Fig. 2-11a. Pertinent points are:

1. Since most valves are pneumatic, a current-to-pneumatic (I/P) transducer is required.
2. The output of the D/A converter can be displayed so that the operator can readily ascertain the valve position.
3. As the output is the actual valve position, some mechanism must be provided so that the computer can read the initial valve position.

Alternatively, a pulse generator can be used in the configuration in

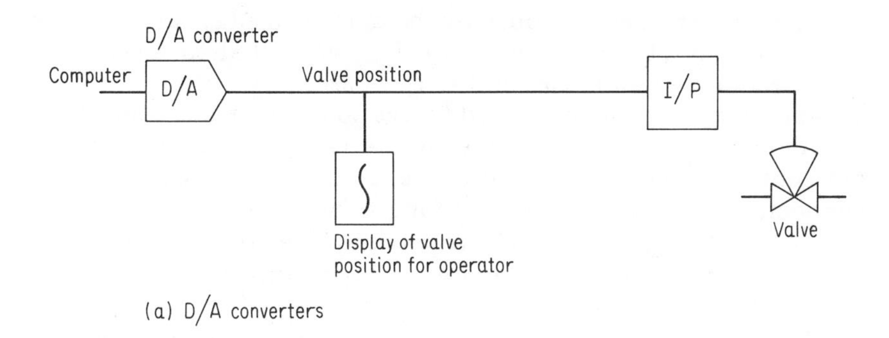

(a) D/A converters

(b) Pulse generator with stepper motor

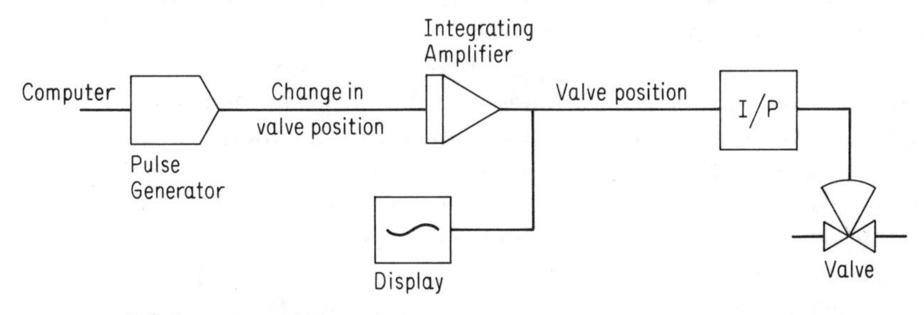

(c) Pulse generator with integrating amplifier

FIG. 2-11. Uses of D/A converters and pulse generators.

Fig. 2-11b. Relevant points are:

1. Current/pneumatic transducer is replaced by the stepper motor, which inherently integrates the input.
2. As the computer output is a change in valve position, it is not necessary that the computer be able to ascertain the valve position (unless it must be verified that the desired change was actually made).

3. In order that the operator be able to readily ascertain the valve position, a signal must be transmitted to the control room, thus entailing another signal lead.

This approach is commonly used for set points of analog controllers.

Another alternative is to use an integrating amplifier located in the control room as illustrated in Fig. 2-11c. This is similar to the configuration in Fig. 2-11b except for the following:

1. The valve position can be readily displayed to the operator.
2. An I/P transducer is required, although this could be incorporated into the integrating amplifier.
3. The saturation limit of the integrating amplifier may not exactly correspond to the valve full-open or full-closed, which may present some problems when using the velocity control algorithm.

Although a pulse generator is illustrated in Fig. 2-11c, a contact closure maintained for variable duration could be used instead.

2.9 SOFTWARE

In one sense, the computer control system can be considered as composed of two classes of elements. The first of these is called the *hardware*, which has been described up to this point. The second is the *software*, which can be defined as everything over and above the hardware required in order for the computer control system to function. This is perhaps the most encompassing definition, more restrictive definitions being available.

Basically there are two sources of software. The computer manufacturer generally supplies certain program packages with the computer system. Some of these are generally included in the basic price of the system. Others may be purchased at the option of the user. In either case, this software is termed *vendor-supplied software.*

Whereas the vendor-supplied software is generally usable in a relatively wide class of applications, each user will require certain programs specifically for his own installation. He has the option of either writing them himself or retaining an outside firm to write them for a negotiated fee. Software in this category is generally termed *user-supplied software*. Of course, the user would like to minimize the amount of software he must develop.

For a computer control system, the software required can be categorized as follows:

1. The operating system, monitor, or executive. This software supervises or directs the operation of the computer control system, scheduling programs for execution, transferring pro-

grams from disk to core, etc. This package is generally available from the computer manufacturer.

2. Supporting software packages, including compilers, loaders, disk editors, diagnostic routines, etc. Most of these are available from the vendor.
3. Applications programs (i.e., those directly concerned with implementing the selected control strategy). Most of these must be supplied by the user, although some parts such as operator's console service routines, thermocouple conversion routines, and the like may be available from the vendor.

With this overview of the software for a control computer, a few of the individual elements will be considered more closely.

2-10 THE ASSEMBLER

Earlier in this chapter we discussed the basic machine language, and indicated how certain operations could be obtained with the appropriate instructions. At this level programming is very tedious, and the programmer must remember the binary codes for each instruction as well as the addresses where each piece of information is stored. Programming in assembly language offers two advantages:

1. Mnemonics are used to indicate the instruction to be performed. For example, STW may indicate the "store word" instruction.
2. Variables are used in the place of absolute storage locations. The assembler collects the names of all variables used in the program and assigns storage locations to them, in much the same manner as the well-known Fortran compiler.

For example, the instruction

```
STW X
```

may instruct the machine to store the contents of the accumulator in the storage location corresponding to variable X.

In most basic assemblers, there is a one-to-one correspondence between assembly language statements and machine language instructions. These assemblers will frequently run on as small a system as one with only 4K words of core. Many manufacturers offer an advanced assembler which permits the use of "macros," which are certain assembly language statements or "instructions" that require the execution of more than one machine language instruction. A more expanded system is generally required for assemblers of this type.

As this is not a text on programming per se, we will not delve into the details of assembly language programming. Besides, an assembly language is generally specific to a specific machine, differing from one model to another even if made by the same manufacturer.

We shall defer discussion of the advantages and disadvantages of programming in assembly language until after our introduction to Fortran.

2-11 PROBLEM-ORIENTED LANGUAGES

Although modifications of other problem-oriented languages such as BASIC have been used for programming process control computers, Fortran is currently the most common problem-oriented language used in process control. As we shall enumerate shortly, the fact that Fortran has its shortcomings has given rise to some interest to abandoning the use of Fortran in process control. However, there is considerable inertia in the general use of Fortran, probably because most current technical graduates have been exposed to it. Consequently, we shall base our discussion in this section around Fortran, pointing out its advantages and limitations.

Fortran entirely abandons the one-to-one correspondence of statements to machine language instructions. Instead, syntax is used to indicate procedures to be executed using desired information. For example, the statement

$$C = A + B$$

indicates that A is to be added to B and the results stored in C. This statement would be equivalent to the following assembly language statements:

```
LDW  B   (load B into the accumulator)
ADD  A   (add A to the contents of the accumulator)
STW  C   (store contents of the accumulator in C)
```

As most readers are certainly familiar with Fortran, there is no need to go into the details of Fortran programming.

It should, however, be pointed out that the Fortran available on process control computers does not generally have as many features available as the Fortran available on the typical data processing machine. Notable exceptions include the absence of logical variables (including logical IF) and the ability to selectively define the precision used in the calculations. In general, single precision or double precision is used throughout the program, not just in selected places where it is needed. The same applies to variables. For example, all integer variables are stored one per word or all are stored one per two words (double precision).

Since the Fortran available on process control computers is really just a carryover of the Fortran available on data-processing machines, several needed features are not generally available. A prime example is the ability to directly perform bit manipulations. The status of process equipment is often indicated by the state of contact closures, which are read into the machine one word at a time. That is, in a 16-bit machine, the status of 16 contact closures would be indicated by one word. Therefore it is often necessary to determine if a certain bit is "on" or "off." This is readily accomplished in assembly language, but not in Fortran.

To provide capabilities like this, the usual approach has been to resort to subroutine calls to assembly-language subprograms that perform the needed manipulations. In addition to bit manipulations, most real-time functions such as initiating A/D conversions, initiating D/A conversions, generating pulse outputs, and the like are handled in this manner. This results in a certain amount of overhead in transferring control to and from the subprogram.

One approach to circumvent this drawback is to permit the insertion of assembly statements into a Fortran source program, a feature called *in-line assembly*. Now the programmer has direct access to the basic machine capability whenever the needed operation cannot be readily accomplished with Fortran.

Basically, the decision to use assembly or Fortran involves a decision of which resource is scarcer—man hours or machine capacity. Certainly a Fortran program can be prepared quicker than can an assembly-language program. However, the assembly-language program will run faster and will require less core storage. Thus a somewhat larger machine will usually be needed in order that the bulk of the programming can be done in Fortran. Other pertinent factors are outlined in Table 2-4.

2-12 FILL-IN-THE-FORMS SYSTEMS

Whether using assembly language or Fortran, the programming burden on the user is substantial. One approach to reducing this burden on the user is via fill-in-the-forms packages, where the user designates by data cards what functions are to be performed. In essence, the master program makes available to the user a number of functions. Via the input data deck, he prescribes what operations are to be performed on designated inputs to produce designated outputs.

TABLE 2-4
Assembly Versus Compiler Languages
(Reproduced by permission from Ref. 5)

Language	Advantage	Explanation
Assembly	Fast object code	Fewer instructions to convert into machine code decreases execution time
	Efficient memory utilization	Assembly code can take advantage of memory-conserving features of modern control computers
	Control over program and data location	Assembly code offers more flexibility in specifying program layout and data storage
	Access to all computer functions and instructions	Programmer can take advantage of his detailed computer knowledge to write more effective control programs
	Efficient program linkage	Calling up subroutines and shifting control parameters is simpler
	Ability to use different classes of codes	Reentrant routines for servicing priority interrupt are facilitated
Compiler	Machine independent and standardized	A limited advantage
	Self-documenting	Yes, but must be supplemented
	Easier to learn	Yes, for a scientist or engineer
	Quicker, less tedious to write or modify	Yes, provided the program writer knows when to provide control alternatives
	Easier to debug—self-checking	Prevents some programmer errors

The functions normally covered by languages of this type include the input scan routine, alarm scanning, conversion of input data to engineering units, three-mode control calculations, feedforward control calculations, cascade control, and similar functions. In general, all of the basic functions common to most control systems are provided.

The fill-in-the-forms system runs in what is called the *interpretive mode.* The input data is stored somewhere in the system, and the fill-in-the-forms system searches through the input data to ascertain what functions are to be performed. This entails considerable overhead as compared to either assembly or Fortran programs written to accomplish the same task, which necessitates a more expanded computer system to perform the same task. The fill-in-the-forms language is not a compiler.

As it is unreasonable to expect any fill-in-the-forms system to provide all the functions required of a computer control system, provision is generally made in these systems for the user to add routines as necessary to augment the system.

2-13 DOCUMENTATION

No matter which programming language is used, preparation of adequate documentation requires considerable effort, but is a task that must be undertaken while preparing the programs themselves. That is, it is not feasible to delay preparation of documentation until the programming task is completed.

In preparing documentation, the objective should be to enable someone who is totally unfamiliar with the program to quickly and easily understand its purpose and how it works. The following items are essential:

1. A written statement of the function this program is to perform, as well as the details of the approach used to accomplish the desired function.
2. A flowchart of the program.
3. An up-to-date program listing.
4. Definitions of all variables used in the program.

This should be augmented as necessary to provide complete coverage.

In regard to defining variables, a standard naming convention for all variables in the various programs in the system has some merits. In this approach, each character in the variable name designates something about its meaning. This method should lead to more consistent variable naming, but a list of variable definitions for each program is still desirable.

2-14 FOREGROUND/BACKGROUND OPERATION

The programs executed by a typical process computer are generally divided into two types: *foreground tasks* and *background tasks.* The foreground tasks are generally those directly involved in controlling the process. The background tasks include many of the tasks required to support the computer control system. For example, the program used to compile programs, whether foreground or background, is run as a background task. Therefore if it is desirable to be able to compile while the computer is on-line (i.e., controlling the process), the operating system or executive must be capable of simultaneously supporting foreground and background tasks on the same machine. This does not mean that both tasks are executed simultaneously. Instead, the background task is executed only while the computer has no foreground task to perform. Systems that cannot support background tasks while on-line are commonly referred to as *dedicated systems.*

Basically, three different arrangements could be proposed to accomplish all the tasks necessary in the operation and support of a computer control system:

1. *Foreground/background on the same machine.* This dual function places an added burden on the monitor or executive system, thereby increasing the overhead. A rather expanded system is required to support both functions. Also, some consideration must be given to the possibility of a background program going astray and interferring with the control programs operating in the foreground. This requires some form of protection, either hardware or software.
2. *Foreground/background on separate machines.* If two machines are purchased from the same manufacturer, one can be dedicated to the control functions while the second is used off-line for program development. Each of the two systems will be of smaller configurations than a machine on which both functions are implemented. Separation of the two functions also eliminates the possibility of background programs interfering with foreground programs. Since one background computer can support several dedicated control computers, this approach can be very attractive when more than one control computer is involved.
3. *Off-line support by data-processing machines.* The general idea behind this approach is that a central computer or timesharing system can be used to provide Fortran compilations and similar functions. As the central computer is most likely of a different make than the control computer, its compilers are not usable. Instead, a compiler is required that runs on the central computer yet produces code executable by the control computer. In the case of Fortran, this removes the restrictions on core made available by the control system to the compiler, and could conceivably permit the development of more efficient compilers that also provide some extra functions needed in process control. As for assemblers, an assembler written in Fortran could be run on virtually any data-processing machine.

2-15 INTERRUPTS

The purpose of an interrupt is to permit the normal flow of execution of instructions to be altered to permit the computer to attend to some urgent or higher priority function. Interrupts are basically of three types:

1. *System interrupts.* These interrupts originate within the computer system itself and play an integral part of the functioning of the system. An example is where the output typer signals the system that it has finished typing the previous character and is ready for another.
2. *Timer interrupts.* These synchronize the operations of the system with the real world. Timer interrupts are generated at prescribed intervals of time, and their occurrence can be used to initiate the execution of control programs (such as algorithm calculations) at regular intervals of time.
3. *Process interrupts.* These originate from the process and either signal alarm conditions, request that some function be performed by the computer, indicate completion of some task within the process, or similar purpose. For example, a high-pressure limit switch could be tied into the interrupt system to indicate alarm conditions in some part of the process equipment. The "request" button on the operator's console is tied to the interrupt system, thereby permitting him to request the computer to perform certain functions. On-stream analyzers often indicate completion of the analysis via an interrupt.

In most process control systems the interrupts play a most important role in the operation of the system.

The interrupt structure varies considerably from one computer system to the next. The sequence of events associated with the occurrence of an interrupt is typically as follows.

1. The interrupt occurs.
2. Instead of executing the very next instruction in sequence, control is transferred to a designated location in core storage and the instruction contained therein is executed. If the interrupt can be serviced by this one instruction, control then reverts back to the program being executed at the time the interrupt occurred.
3. If execution of several instructions is required, the instruction executed due to the interrupt is generally a special instruction that stores the current contents of the address register and loads into the address register the location of the next instruction to be executed.
4. The instruction located at the address now in the address register is the first instruction in a program called the *interrupt service routine.* However, the information in the working registers pertains to the program in execution when

the interrupt occurred. In order to resume execution of that program, their contents must be stored. The initial instructions of the interrupt service routine must accomplish this task.

5. The instructions to accomplish the function relative to servicing the interrupt are executed.
6. The contents of the working registers are restored to their values at the time the interrupt occurred.
7. The contents of the address register is restored to its value at the time the interrupt occurred.

After the last step, the program in progress when the interrupt occurred is resumed from the point at which it was interrupted.

Process computers come in several different "styles" in regard to their interrupt structure. In one style, there is essentially only one interrupt priority. Upon initiation of the servicing of any interrupt, all other interrupts are "inhibited" (i.e., servicing is not permitted until the one currently being processed is completed). In this type of system the interrupt service routines must generally be short.

In another variation, interrupts are grouped into levels of different priority, with several interrupts being tied into each level. In this system, interrupts occurring on high-priority levels will interrupt the servicing of interrupts on lower-priority levels. However, an interrupt will not interrupt the servicing of another interrupt on the same level.

In yet another variation each interrupt is provided its own distinct priority, and interrupts the servicing of interrupts of lower priority.

Some degree of program control is provided by inhibit commands which prohibit the recognition of all or selected interrupts until the machine is returned, under program control, to the uninhibited state.

2-16 THE EXECUTIVE

The operation of the process control computer is under the supervision of the executive, which is alternatively referred to as the *operating system* or *monitor*. One of its primary functions is to schedule the execution of control programs. Somewhere within the system is located all control programs which can be executed by the computer. Some of these may be located in core at all times, and are termed *core resident*. Others may be located on the disk or drum, if available. In this case, the monitor must supervise the transfer of the programs from the disk or drum to core storage.

The scheduling of execution of control programs is accomplished with the aid of a table called QUEUE, which contains the name of all programs whose execution has been requested but not fulfilled. Along with each program is an associated priority, which is assigned under program control at the time the program's name is placed in QUEUE. Program names are placed into QUEUE mainly by one of the following ways:

1. A control program may place the name of another program into QUEUE, thereby permitting a train of successive programs to accomplish a given task rather than one large program.
2. An interrupt service routine may place the name of a program into QUEUE. In many cases, this is the only function of the interrupt service routine.

Once a program's name is placed into QUEUE, it is removed only when the program is executed. Highest-priority programs are executed before low-priority programs. Programs having the same priority are executed on a first-in, first-out basis.

On systems operating with a disk or drum, the layout of core storage is as illustrated in Fig. 2-12. The executive generally resides

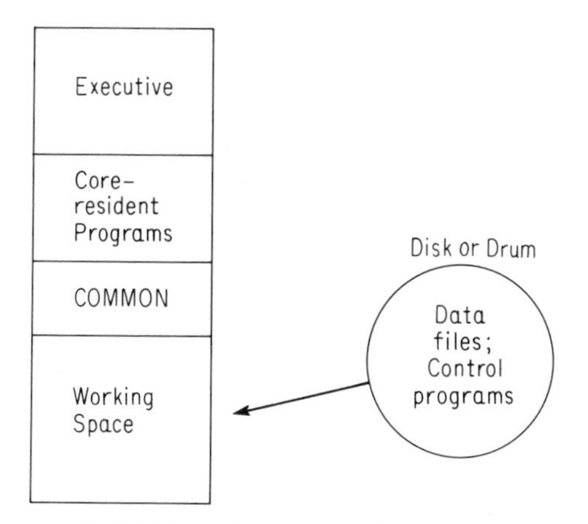

FIG. 2-12. Disk-oriented operating system.

in the lower portion of core storage. An area of core storage called COMMON is reserved for the storage of frequently used data. Core-resident routines remain permanently in core storage. The remainder of core storage is called *working core*. It is into this area that the programs residing on the disk are loaded for execution.

When the execution of a control program residing on disk is scheduled, the program is copied from the disk into working core by the executive. Note the word copied—the original version on the disk is not altered. After execution of this program has been completed, the next program is copied into the same area of working core (i.e., it overlaps the original program). This means that the program is not returned to disk after execution is completed. Therefore, the same program is executed each time, only the data being different. Since the completed program is not copied back onto the disk, any data that may be needed next time the program is to be executed must be stored either in COMMON or in a file on the disk.

Depending upon the executive, the working core area may contain either only one program at a time, a specified maximum number of programs, or however many can be accommodated in the space available. In systems that can accommodate only one program at any given time in working core, the procedure is as follows:

1. QUEUE is consulted to determine which control program is to be executed.
2. The control program is loaded.
3. The control program is executed.
4. Return to step 1.

That is, QUEUE is consulted only at the completion of execution of a program. But as interrupts can be serviced while the control program is being executed, it is conceivable that an interrupt service routine could place the name of a control program into QUEUE whose priority exceeds that of the program now being executed. In most cases this program would not be loaded until execution of the program currently in working core has been completed.

The capability of multiple programs residing in working core storage at any one time is referred to as *multiprogramming*. When these programs may reside only at certain locations in core, this operation is said to be using fixed partitions, as illustrated in Fig. 2-13. Control programs are generally assigned to a particular partition and will only be executed in this partition. The term *dynamic storage allocation* is applied to the case when the program may reside in any area of working core. As illustrated in Fig. 2-13, this leads to a more efficient utilization of working core, but is more demanding on the executive and also requires some supporting hardware features (program location register) in the CPU for efficient implementation.

Although more than one program may reside in working core at any one time, only one program is actually being executed. The others are said to be in the *suspended state*.

Multiprogramming systems generally check QUEUE both upon completing execution of a control program and upon completion of

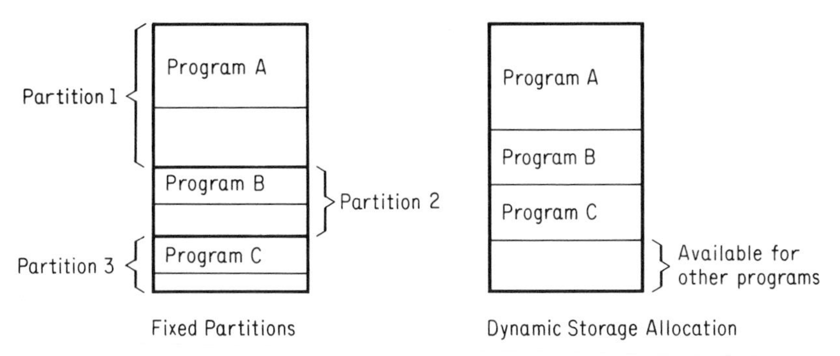

FIG. 2-13. Core allocation in multiprogramming systems.

an interrupt service routine. Thus, if a high-priority program has been entered into QUEUE, the executive loads it for execution provided space is available. In fixed partition systems, this generally means if the partition assigned to the program to be executed is not currently in use. For executives using dynamic storage allocation, this means if the unused area of working core is large enough to accommodate the program.

Some systems using dynamic storage allocation will remove low-priority programs to make room for high-priority ones. This is a rather ambitious undertaking. One approach is to not remove the program, but to store on disk the address in the program at which execution was terminated, the contents of the working registers, and the current values of all data used in the program. The program itself is then overlaid. When space is available for resumption of execution, a fresh version of the program is copied into working core, the working registers and data values are restored, and execution resumes.

2-17 FIRMWARE

The executives described in the previous section have one property in common—they all contain "bugs." Even with considerable effort on the part of both vendor and user, a few bugs still show up from time to time. In addition, the software executives also entail a certain amount of computational overhead to perform the desired duties. The executives also require considerable core storage, often as much as 50 percent of the available core.

One approach to circumventing these drawbacks is via a firmware executive, i.e., one that is hardware-implemented rather than software-implemented. This approach, however, has the disadvantage of generally being inflexible.

One approach is to essentially organize the entire control programs in to a small number, e.g., eight, of rather large packages. The computer executes one package for a fixed length of time, then proceeds to the next. Thus each program receives a fixed percentage of the machine's time during each cycle. For example, one program might be the service routine for the operator's console. The program then loops on an instruction that determines if the operator has requested anything. When the operator enters a request, the program determines what was requested, performs the requested function, and returns to the loop to await entry of another request.

LITERATURE CITED

1. Butler, J. L., "Comparative Criteria for Minicomputers," *Instrumentation Technology*, Vol. 17, No. 10 (October 1970), pp. 67-82.
2. Copeland, J. R., and S. P. Jackson, "Minicomputers for the Control System," *Control Engineering*, Vol. 16, No. 8 (August 1969), pp. 90-94.
3. Lawrence, J., "Computer Input Noise—Avoid it, Don't Fight It," Proceedings of Second Annual Conference on the Use of Digital Computers in Process Control, Louisiana State University, Baton Rouge, La., Mar. 8-10, 1967.
4. Lewtas, R., J. Taylor, and J. B. Neblett, "Executive Software," Proceedings of the Third Annual Workshop on the Use of Digital Computers in Process Control, Louisiana State University, Baton Rouge, La., Feb. 21-23, 1968.
5. Kipiniak, W., and P. Quint, "Assembly vs Compiler Languages," *Control Engineering*, Vol. 15, No. 2 (February 1968), pp. 93-98.

chapter 3

Supervisory Computer Control

Perhaps the most commonly used computer control scheme in the process industry today is the supervisory control concept introduced in Chapter 1. That is, the usual plant analog control systems (or a separate DDC computer) are responsible for the first-level control functions, the supervisory computer simply prescribing the proper operating instructions (e.g., set points) so that the desired objectives are met to the best degree possible. This chapter describes the two basic mathematical prerequisites for implementing such a control scheme: 1) developing a plant model, and 2) optimizing the operating conditions.

Probably the best way to study the subject of modeling is via case studies. As modeling is a subject appropriate to a book in itself, we must of necessity limit the scope of our discussions. The main objective of the ensuing discussion is to show the role of models and optimization techniques in supervisory control systems, pointing out a few aspects that are either unique or extremely important in their application.

3-1 CLASSES OF MODELS (1, 2, 3)

All supervisory control systems are based on models. In fact, the general inavailability of the necessary models is probably the most difficult hurdle to overcome in justifying a supervisory system. The cost of the model development effort is usually a substantial portion of the total project cost, and is often difficult to estimate in advance.

Most supervisory systems contain at least one use of each of the following types of models.

Physical Model

Being the objective of most modeling ventures, the physical model consists of mathematical relationships between the various operating variables. This model encompasses the various transfer mechanisms, kinetic relationships, and conservation laws commonly resorted to in order to develop a mathematical description of the unit. A number of factors must be considered when developing the physical model. For example, should the relationships be dynamic or steady-state? Should analytic or empirical approaches be included? What is the objective of the model? How much of the plant should it describe? Most of these points will be considered in more detail in later sections.

Procedural Model

Although the physical model is perhaps more elegant, our ability to mathematically describe our processes has its shortcomings, forcing us to consider other approaches. To a large extent, the procedural model is just as implied by the name—a procedure for accomplishing some objective. For example, to start a turbine in a power generating station, a prescribed sequence of actions comprise the procedural model for accomplishing this objective.

In many cases procedural models are used to instruct the computer to duplicate the same functions that would otherwise be accomplished by the operator. Although this approach does not inherently offer any improvement over the operator, it will surpass him in consistency. With this assurance that the same procedure is used in every instance, it is frequently possible to accurately evaluate various alternatives or polish some of the fine points in the procedure.

Another area in which procedural models are important is safety. If the process conditions evolve into a dangerous situation, the computer may be instructed to abort normal functions and follow some preassigned instructions of a procedural model to insure that the plant is quickly returned to a safe region of operation. Clearly, a physical model is not normally needed in this kind of application.

To illustrate the fact that either a procedural model or a physical model may be used for certain applications, consider a grade change on a Fourdrinier paper machine. That is, the machine is operating in near-steady-state conditions so as to produce paper of a certain grade, weight, strength, and moisture. It is now desired to change the grade. Normally the optimum approach is to make this change in the minimum possible time, as in many cases the paper produced in the interim period cannot be sold as product. In noncomputerized mills, the operators usually have some set procedure that has evolved over the years for accomplishing this objective. This procedure comprises the *procedural model.*

An alternative to this approach is to develop a dynamic physical model of the paper machine. The grade change is nothing more than the transfer of the system (i.e., the paper machine) from one operating point or state to another. This is mathematically very similar to the missile trajectory problem (i.e., transfer of a system from one state to another), to which optimal control theory has been applied to determine the optimal trajectory. If a dynamic model of the paper machine were available, these techniques could be applied to minimize the time required for a grade change. This is not normally undertaken, primarily because 1) the computational solution of the optimal control problem is not simple, 2) the economic incentives are not very great, and 3) mathematical models of sufficient accuracy are not generally available. The improvement over the operator's procedure will not justify the extra effort.

Although rarely mathematically rigorous and frequently quite crude, procedural models provide an important basis for many of the functions of the supervisory control computer. It is in this area that a good working relationship between the plant operating personnel and the computer control personnel can pay good dividends.

Economic Models

Economic models differ from physical models in that they mathematically relate economic factors rather than process variables. The basic objective of plant operation, namely, to maximize profit or minimize cost, can be applied most directly to models in this category.

The culmination of any model is the calculation of the objective (or criterion) function, a single number which measures the performance of the process. The optimum process operation is that which extremizes (maximizes or minimizes) the objective function. The natural objective function to propose for most industrial processes is the profit contribution, given by

$$\text{P.C.} = \sum_{i=1}^{n} V_{P_i} P_i - \sum_{j=1}^{m} V_{M_j} M_j$$

where

V_{P_i} = value of one unit of ith product
V_{M_j} = value of one unit of jth raw material
P_i = production rate of ith product
M_j = consumption rate of jth raw material

It should be noted that V_{P_i}, although not appearing explicitly, may be a function of the quality variable for the ith product.

Although the profit function appears straightforward, its direct implementation is beset with serious difficulties. Many processes are operated within a large complex, most of its raw materials coming from its neighbors and most of its products going to its neighbors. In such situations the sales price in the competitive market may not really be appropriate. Although the plant accounting department uses some values for their internal purposes, these often are not realistic for some reason or other. This is even further complicated by the fact that what is best for this particular plant may not be best for the corporation as a whole.

In many situations, production is limited by the capacity of the plant, not by the market situation. If either (1) the cost to produce a unit of product is independent of the rate of its production or (2) the product's value is much higher than the cost to produce it, the profit function normally reduces to maximum production. This is generally the case for operations such as paper mills or ammonia plants. If such cases can be recognized early in the development of a model, considerable simplifications are possible, resulting in savings in time and effort. Specifically, most of the terms in the profit function are not needed, and accurate values of the economic variables are unnecessary.

Perhaps the most difficult situation is one in which the plant must produce a specified quantity of product at the minimum cost. This fixes the largest term, the return from sales, in the profit function. Smaller magnitude terms become important, most of which could otherwise be neglected. There is generally a rather large number of these, and discretion must be used to ascertain which ones need to be included in the analysis. Even so, the resulting profit function along with the mathematical model to relate production and quality variables is likely to be more complex.

3-2 CONSTRAINTS

Thus far we have studied to some extent two parts of a complete mathematical model for a plant. Specifically, we have considered the objective function to measure the performance, and the physical model that relates the performance variables in the objective function. The third part of the complete model is equally as important, being the constraints within which the process must operate. These constraints arise from physical limitations, such as safety considerations and raw-material availability.

The maximization or minimization of the objective function now becomes a constrained optimization problem. Assuming constant product and raw-material values, the profit function is linear in the product and raw-material quantities. Although nonlinearities arise from the dependence of the product rates upon the raw-material rates and other parameters (as per the physical model), the degree of nonlinearity is often not too great. From linear programming theory, it is known that the optimization of a linear function will always give an answer lying on the boundaries given by the constraints. This will also generally hold when only a slight degree of nonlinearity is imparted by the physical process model, although it is possible for there to be a true "peak" within the boundaries of the constraints.

From the above discussion it should be obvious that the constraints are extremely important. Not only are accurate limits needed, but there must also be the proper number. Too few constraints will produce an unreasonable result, frequently being infinite levels of operation. Although too many constraints do not affect the result (as long as the constraints are correct), there is no return from the time and effort invested in their formulation and incorporation into the control strategy.

The general form of constraints may take one of two forms. They may be equality constraints of the type

$$P_i = \text{fixed value}$$

which says that the production rate of the ith product is fixed. Alternatively, they may be inequality constraints of the type

$$\text{Lower limit} \leqslant P_i \leqslant \text{upper limit}$$

which restrict the production rate of the ith product to a specified range.

Physically, the constraints may arise from several sources. The use of good engineering judgment is perhaps the best method for their formulation. They may arise from such areas as safety (pressure in vessels cannot exceed their design rating), quality (product must be salable), equipment capacities (overload of equipment generally results in excessive maintenance), process characteristics (limiting reactor temperature to minimize production of byproducts), and other similar sources.

It is not always desirable to incorporate the same constraints as used by the operators. Examples could be in compressor operation, where constraints on the prime-mover speed and horsepower might be more appropriate that the limits on flow rates and suction pressures normally used by the operators.

3-3 MODELS FOR SUPERVISORY COMPUTER CONTROL

To a large extent, the primary deterrent to the application of supervisory control is that the necessary models are not available. Usually the only model available is the one used at the design stage. Although some useful relationships can be realized from this model, it should be recognized that the objective of the design model is different from that of the control model. The design model is used to calculate the physical characteristics of plant equipment from specified operating levels. On the other hand, the control model must relate the operating variables for a plant whose physical characteristics are known.

Although supervisory control is usually presented as optimizing the steady-state operation of a process, the same concepts can be applied to batch processes, an example being gasoline blending. Some processes are "batch" at the reaction stage and "continuous" elsewhere. If the plant is reaction-limited, it then becomes desirable to optimize the reaction cycle. The objective is to select a reaction time and temperature/pressure program that optimizes the production of valuable products. Since the reactor is batch, an unsteady-state model is required. By our terminology, the control system would still be supervisory in nature if the computer provided its results to the first-level control system (whether analog or digital) instead of implementing them directly in the manipulated variables.

Even in cases in which the process operates virtually at steady state, some dynamic considerations frequently enter. An example could be in catalytic reaction systems, where carbon deposits on the catalyst or some form of catalyst poisoning is a function of the flow rate, temperature, pressure, or other reaction conditions. Here the rate of accumulation of the foreign material upon the catalyst, a dynamic consideration, becomes a very important factor, as increased build-ups lead to shorter times between catalyst regeneration or replacement.

The above two paragraphs have probably emphasized the exceptions, although their importance should not be dismissed as inconsequential. Most plants operate in hopefully narrow limits about some steady-state conditions, and it is the adjustment of these operating objectives toward more optimum values that generates the return for a supervisory system. For these purposes, a steady-state model is quite satisfactory. Steady-state models are likely to contain few if any of the differential equations that usually complicate the numerical calculations associated with the use of a model. Steady-state models are usually much simpler to formulate than their dynamic counterparts, giving another incentive for avoiding the unsteady-state model if possible.

No matter what avenue is selected, it should be recognized that developing a model for a plant is a major undertaking. The various approaches are generally classified into theoretical (analytic) as opposed to empirical techniques. The decision as to which one to use is not made for the plant as a whole, but relegated to each individual component. Even at this level it is frequently not obvious which approach will be more successful. The total effort will always be best achieved by a combination of the two, and the practitioner well versed in both will have a distinct advantage over the expert in one approach alone.

The theoretical approach is expounded in several texts on model development (4–7) and will not be treated in detail here. Texts on unit operations often offer additional aid for particular cases (8) as well as texts on kinetics (9) and the transport sciences (10). Aspects of the model such as heat and material balances can almost always be arrived at theoretically without a large investment in effort. Certainly it is ridiculous to consider an empirical relationship for these. In many cases it is surprising what simple heat and material balances alone can show about the process operation. Beyond these, however, the theoretical approach runs into more difficulties.

A common approach to model development is to begin by pursuing the theoretical approach until all the straightforward relationships have been incorporated. At this point there will usually be several important relationships yet to be formulated. It is here that the empirical approach versus the theoretical approach must be examined. For example, suppose that a chemical reactor must be modeled. The exact reaction mechanisms may not be known, and even so, numerical values for the rate expressions are rarely known. In such cases experimental data must be taken no matter how the model is formulated.

Two approaches are open. First, one may write the kinetic expression for a proposed reaction mechanism and determine the numerical values that represent the experimental data in some "best" sense. This we shall call the *theoretical approach*. The alternative is to use regression techniques to relate the observed experimental data. This we shall call the *empirical approach*.

There are obvious advantages and disadvantages to both approaches. The theoretical approach will often require substantially more effort. Several alternatives must be evaluated, especially where reaction mechanisms are unknown, with the finite possibility that none will be completely satisfactory. A substantial amount of computer effort may be required, along with controlled plant tests or laboratory data to verify the mechanism.

As it rests on a sounder basis, it would be expected that the theoretical model would be more general than its empirical counterpart. This is especially true when it becomes necessary to extrapolate the results beyond the region in which the experimental data were taken.

The theoretical model is likely to contain more cumbersome relationships than those resulting from a regression analysis. It may contain complex nonlinear relationships or differential equations, perhaps even partial differential equations. It is essential that these relationships should not become so complex that the computational capabilities of the digital computer are exceeded.

In considering the theoretical approach, one must be sure that one is not undertaking a long-term research project. In such cases the theoretical approach should be recommended to the research department, allowing the computer control project to proceed by using the more empirical approach.

Of course, the theoretical aspects of many operations in the process industries are simply not understood well enough to allow a theoretical analysis. A common suggestion for such cases is to correlate past plant records. These data are usually recorded in a narrow region about the desired operating levels. Thus a major portion of the data is effectively identical. During the relatively brief periods in which the process is removed from these levels there is always the tendency either to not record the data or to slant it toward more desirable values.

A more serious problem encountered in using this data is that the proper cause-and-effect relationships cannot always be determined. For example, suppose variable A is increasing. The operator then consistently makes an adjustment that increases the value of variable B. In the regression analysis, it will appear that A and B are strongly correlated when in fact there is no correlation at all.

No matter which approach is taken, the final results will be no better than the data upon which they are based. The preferred route is to use closely supervised plant tests while employing the most accurate transducers available. Some companies have gone so far as to construct special-purpose vans to facilitate making such tests.

The literature contains discussions and developments of process models that have been used in supervisory control systems or are suitable for such use for ethylene plants and/or cracking furnaces (11-14), catalytic cracking units (15), methanol plants (16), ammonia plants (17), and others. Most of these models either permit or require a certain amount of tailoring to fit a particular unit, and in many cases certain parameters must be updated and improved as more information about the process becomes available.

3-4 OPTIMIZING CONTROL OF A DISTILLATION COLUMN

The optimizing control strategy for a distillation column presented by Baxley (18) is designed to operate the column in Fig. 3-1 in such a manner that the economic return is optimized. This column

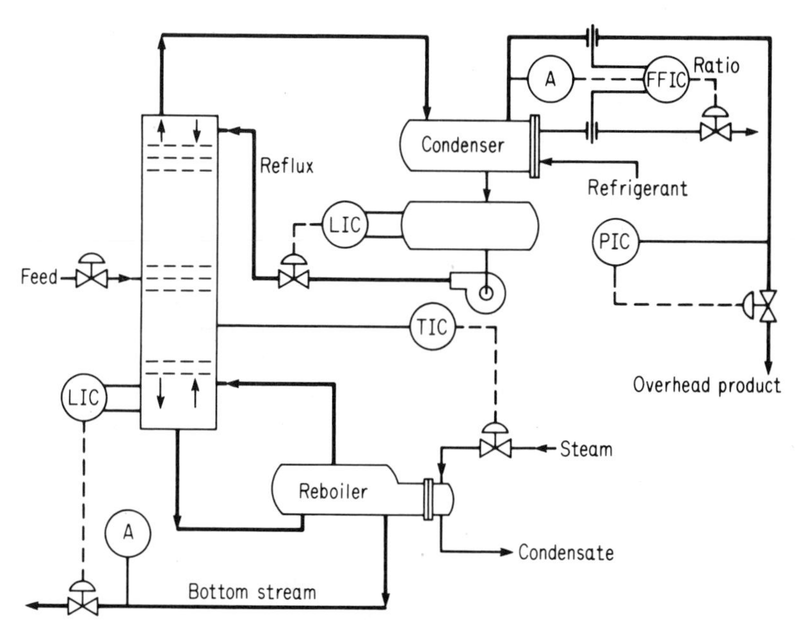

FIG. 3-1. Column with conventional control system. (Reproduced by permission from Ref. 18.)

is one in a series in a separation facility. Tight purity specifications must be met on the overheads, whereas there are no specifications on the bottoms. Therefore there should be an economically optimum amount of light key lost in the bottoms. Refrigerant is used as the cooling medium in the condenser.

The first-level control system consists of the following:

1. Liquid level in the reboiler is maintained by adjusting the bottoms rate.

2. Liquid level in the accumulator is maintained by adjusting the reflux rate.

3. The steam rate (and consequently the boilup rate) is adjusted to maintain a constant temperature on a tray in the stripping section.

4. Constant column pressure is achieved by manipulating the overheads rate.

5. A ratio controller is used to maintain a constant ratio of refrigerant-to-overheads rate.

6. A composition analyzer on the overheads stream adjusts the ratio setting on the ratio controller.

Actually, the optimizing technique to be described below is independent of the first-level control strategy employed. The main requirement is that the control strategy maintain good regulation of the unit, i.e., maintain the controlled variables near their set points. Several discussions of distillation column control schemes are available in the literature.

As long as the overhead product is of the specified purity, it can be sold, i.e., the plant is not operating in a market-limited environment. Therefore, operation of the column is optimized by increasing the production of overhead product until either the optimum recovery or an equipment constraint is reached. The optimum recovery is that point at which the incremental cost of utilities equals the value of the incremental product produced.

In many industrial columns the value of the product is much greater than the cost of the utilities required to produce it. In this case it will always be profitable to increase the recovery until an equipment constraint is reached. In other words, the column should be operated so as to produce maximum recovery, and the economic optimum recovery need not be computed. Probable exceptions are columns for which (1) refrigerant is used in the condenser or (2) operation is at high reflux ratios (i.e., superfractionators).

For the column in Fig. 3-1, three equipment constraints must be considered.

1. *Condenser.* Either (1) the refrigerant valve is fully open and the condenser is heat-transfer limited, or (2) the maximum amount of refrigerant allocated to this column is being consumed.

2. *Reboiler.* Again, the steam valve may be fully open (reboiler is heat-transfer limited) or the allocated amount of steam is being consumed.

3. *Trays.* If the vapor and liquid rates within the column become too high, the column may "flood." To avoid this condition, a limit is placed on the allowable pressure drop from top to bottom of the column.

For distillation columns, each of these constraints can be related to the flow rate of vapor leaving the top tray. In the ensuing paragraphs, we shall relate the economic optimum recovery and the above three constraints to the top-tray vapor rate. In operating the column, the top-tray vapor rate can be gradually increased until either (1) a constraint is reached, or (2) the optimum recovery is attained.

In operating the column, only small changes in the top-tray vapor rate will be made each time the optimizing calculations are made. Therefore the equations need be accurate only in a fairly small region about the current operating point. For this reason, fairly crude approximations to rather complex processes are acceptable.

The flowchart in Fig. 3-2 gives the logic used in the optimizing scheme. We shall consider each block separately.

Calculate Top-Tray Vapor Rate

The top-tray vapor rate (TTV) can either be (1) measured directly or (2) calculated as the sum of the overhead product rate and the reflux rate.

Condenser Constraint

At the constraint, the refrigerant motor valve to the condenser will be fully open, regardless of whether the cause is refrigerant availability or heat-transfer capability. The top-tray vapor at the constraint (VLIM1) can be extrapolated from the current top-tray vapor rate (TTV), the current valve position (PMV), the lower limit on the valve opening (AOLO), and the upper limit (AOHI):

$$\text{VLIM1} = \text{TTV} * \left(\frac{\text{AOHI} - \text{AOLO}}{\text{PMV} - \text{AOLO}} \right) \tag{3-1}$$

Reboiler Constraint

Although the top-tray vapor rate at the reboiler constraint could also be extrapolated from the motor valve position, an alternate approach would be to compute the limit as a function of the heating-medium temperature. The current heat-transfer rate q in the reboiler is

$$q = UA\,(t_s - t_f) \tag{3-2}$$

where U = heat-transfer coefficient
A = heat-transfer area
t_f = fluid temperature (measured)
t_s = temperature of heating medium in the reboiler

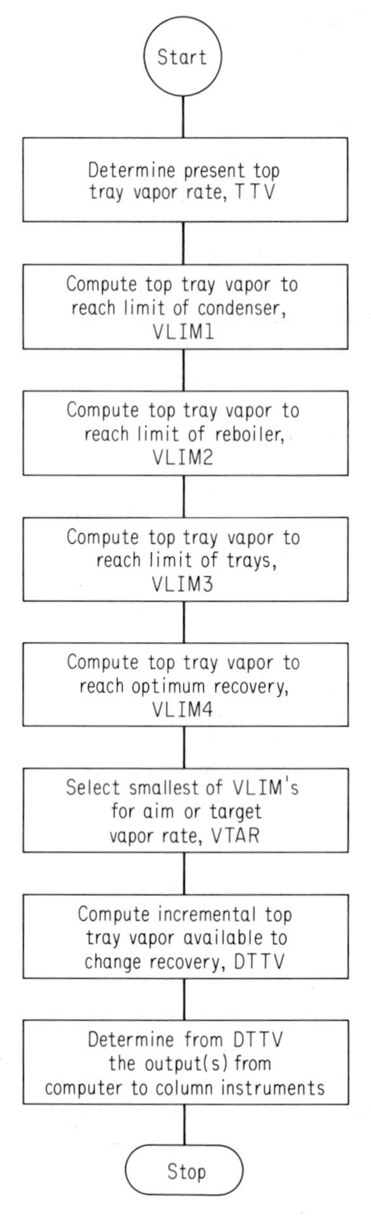

FIG. 3-2. Flowchart of optimizing scheme. (Reproduced by permission from Ref. 18.)

At the limit, the heat-transfer rate q' is given by

$$q' = UA(t_s' - t_f) \tag{3-3}$$

Where t_s' is the maximum temperature of the heating medium. Since the vapor rate is directly proportional to the heat-transfer rate,

$$\frac{\text{VLIM2}}{\text{TTV}} = \frac{q'}{q} = \frac{t_s' - t_f}{t_s - t_f} \tag{3-4}$$

where VLIM2 is the top-tray vapor rate at the reboiler constraint.

When steam is used as the heating medium, t_s and t_s' can be related to the steam pressure.

An analogous approach could be used for the condenser.

Tray Constraint

The top-tray vapor rate at the limit (VLIM3) can be linearly extrapolated from the current top tray vapor rate, the current column pressure drop (PDPRES), and the maximum allowable pressure drop (PDMAX):

$$\text{VLIM3} = \text{TTV}* \left(\frac{\text{PDMAX}}{\text{PDPRES}}\right) \tag{3-5}$$

Alternatively, the extrapolation could be based on the square-root relationship between vapor rate and pressure drop.

Because of gradual solids deposition on the trays, the upper limit on the pressure drop must be updated periodically. Furthermore, in some columns the limit occurs within a small section of trays. In these cases the limiting situation can be more accurately detected from a pressure-drop measurement over this small section of trays.

Optimum Recovery

As defined earlier, the economic optimum recovery is the point at which the value of a small increment of production equals the cost of the increment in utilities required to produce it. If F is the feed rate, x_f the fraction of light key in the feed, and X_R the fraction of light key recovered in the overheads, the distillate rate is given by

$$D = Fx_f X_R \tag{3-6}$$

If u is the value of a unit of distillate, the return from an incremental increase in the distillate rate is

$$u\,dD = uFx_f\,dX_R = u\frac{D}{X_R}\,dX_R \tag{3-7}$$

Since the fraction loss X_L is $1 - X_R$, this expression becomes

$$-u \frac{D}{1 - X_L} dX_L$$

If r is the cost of utilities to produce a unit of reflux, the cost of an incremental increase dR in the external reflux ratio R is

$$rD\,dR$$

Since the incremental return must equal the incremental cost at the optimum recovery, the resulting expression is

$$rD\,dR = -u \frac{D}{1 - X_L} dX_L \tag{3-8}$$

or

$$\frac{dX_L}{dR} = -\frac{r}{u}(1 - X_L) \tag{3-9}$$

This relationship holds only at the optimum.

For a specific column, the fraction loss X_L can be computed for various reflux ratios to give a family of curves such as in Fig. 3-3.

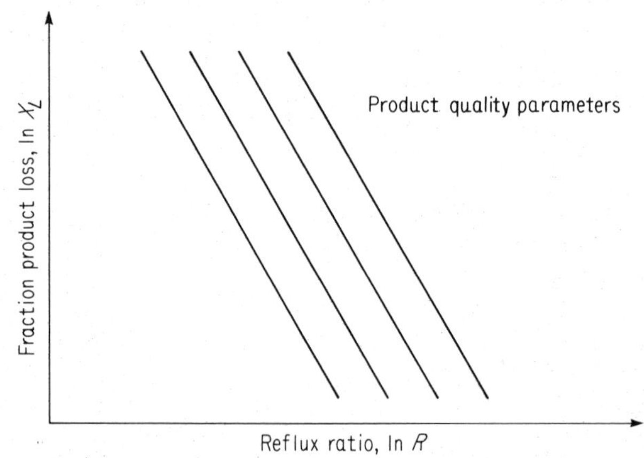

FIG. 3-3. Separation characteristics of the column. (Reproduced by permission from Ref. 18.)

For these curves, the logarithmic relationship is linear in the expected operating range:

$$\ln X_L = k + m \ln R \tag{3-10}$$

or

$$X_L = kR^m \tag{3-11}$$

where k is the intercept and m is the slope of the log-log plot. This relationship should hold both at the current operating point and at the optimum. Using primes to denote the optimum, the fraction loss and reflux ratio at the optimum are related to their current values as follows:

$$\frac{R'}{R} = \left(\frac{X_L}{X'_L}\right)^{-1/m} \tag{3-12}$$

Differentiating Eq. 3-10 gives

$$\left(\frac{dX_L}{X_L}\right) = m\,\frac{dR}{R} \tag{3-13}$$

At the optimum,

$$\frac{dX_L}{dR} = m\,\frac{X'_L}{R'} \tag{3-14}$$

Substituting into Eq. 3-9;

$$\frac{mX'_L}{R'} = -\frac{r}{u}(1 - X'_L) \tag{3-15}$$

Assuming that very small fraction losses occur at the optimum, the $(1 - X'_L)$ term can be neglected to obtain

$$\frac{mX'_L}{R'} = -\frac{r}{u}$$

or

$$R' = -\frac{muX'_L}{r} \tag{3-16}$$

The loss at optimum can be estimated by substituting into Eq. 3-12 and rearranging.

$$X'_L = -\left(\frac{rRX_L^{-1/m}}{mu}\right)^{m/(m-1)} \tag{3-17}$$

Since small steps will be taken in approaching the optimum, the ratio of the top tray vapor at the optimum recovery (VLIM4) to the current top-tray vapor equals the ratio of the reflux ratio at the optimum to the current reflux ratio:

$$\text{VLIM4} = \text{TTV}\left(\frac{R'}{R}\right) = \text{TTV}\left(\frac{X_L}{X'_L}\right)^{-1/m} \tag{3-18}$$

Target Top-Tray Vapor Rate

Of the four limiting vapor rates, the smallest is selected as the target vapor rate (VTAR):

$$\mathrm{VTAR} = \min \{\mathrm{VLIM1}, \mathrm{VLIM2}, \mathrm{VLIM3}, \mathrm{VLIM4}\} \qquad (3\text{-}19)$$

If the result differs too much from the current top-tray vapor rate, VTAR is adjusted so that only small steps are taken.

Set-Point Adjustments

The difference DTTV between the target vapor rate and the current top-tray vapor rate

$$\mathrm{DTTV} = \mathrm{VTAR} - \mathrm{TTV} \qquad (3\text{-}20)$$

can be treated as an error signal for a conventional one-, two-, or three-mode algorithm whose output is the set point of the side-temperature controller in Fig. 3-1. Computing the interaction of this adjustment with the other control loops on the column and compensating accordingly minimizes disturbances in the other loops caused by the change in boil-up rate.

Extensions

Although Baxley did not discuss these, other authors have described further adjustments to increase the recovery from the column. One additional variable could be the column pressure (19). For example, suppose the reboiler is heat-transfer limited and is operating with the steam valve full open. If the column pressure is decreased, the fluid temperature in the reboiler decreases, which increases the temperature driving force for heat transfer and the boilup rate increases. Presumably the column pressure could continue to be lowered until another constraint, e.g., condenser, is reached. The feed enthalpy could also be adjusted to produce somewhat the same results (20).

Baxley briefly discusses a case where one supervisory function must communicate with another. If the refrigeration unit is supplying the cooling medium to several columns of which one or more is being controlled in the manner described previously, the allocation of the refrigerant to the various columns becomes another optimization problem. The results of this calculation would be used in the calculations for the condenser limit as described previously.

3-5 OPTIMIZATION TECHNIQUES

As in the case of modeling, a number of excellent texts are available on optimization techniques (21,22,23), and consequently no detailed discussion of any technique will be presented here. Instead, we will only briefly describe to what extent the various techniques are generally used in supervisory control systems.

Linear programming (LP) finds common use in blending problems. For example, a simplified version of the blending operation for a cement plant (2,24) requires that two materials be combined to meet certain specifications on a feed stream to the plant. The composition and value of the two materials which we shall simply call A and B are

Component	Percent Ca	Percent Mg	$ per pound
A	46	1	.20
B	40	4	.10

Suppose that the following restrictions are placed upon the blending operation:

1. The percent Ca of the blended stream must be between 42 and 44 percent.
2. The percent Mg of the blended stream must be less than 2.6%.

The performance objective for the blending operation is to combine the two raw-material streams in such a manner as to produce a feed stream that meets these requirements at minimum cost.

If we develop our equations based upon one pound of feed material, the performance functional becomes

$$0.20W_A + 0.10W_B = \text{cost}$$

where W_A = lb of A and W_B = lb of B. As one pound of blended material is to be produced, a material balance gives an equality constraint:

$$W_A + W_B = 1$$

The mathematical formulation of the inequality constraints is

$$0.420 \leqslant 0.46W_A + 0.40W_B \leqslant 0.440$$

$$0.01W_A + 0.04W_B \leqslant 0.026$$

Thus we have a performance functional to be minimized subject to one equality constraint and three inequality constraints.

Because all equations are linear, this problem can readily be solved via linear programming. The solution, namely 0.467 lb A and 0.533 lb B to form 1.0 lb product for the above example, is known to lie on the boundary described by the constraints, which makes their proper formulation extremely important. The simplex method used in the solution of the linear programming problem can be shown to converge in a finite number of iterations, a characteristic rarely possessed by optimization techniques.

Another application of linear programming has been to the economic dispatch problem in the power industry (25,26,27).

Although not directly applicable to nonlinear problems, the concept of linear programming can be applied via a technique called *sectional linear programming.* As illustrated by the flowchart in Fig. 3-4, sectional linear programming uses a linear approximation to the nonlinear problem, the linearization being computed at the best current estimate of the optimum. Unlike linear programming, convergence of the sectional linear programming algorithm is not guaranteed. It is subject to cycling when the problem is highly nonlinear. However, it has enjoyed considerable popularity because it has many of the desirable features of linear programming, such as the capability to readily incorporate constraints.

Despite the similarity in their names, *dynamic programming* and

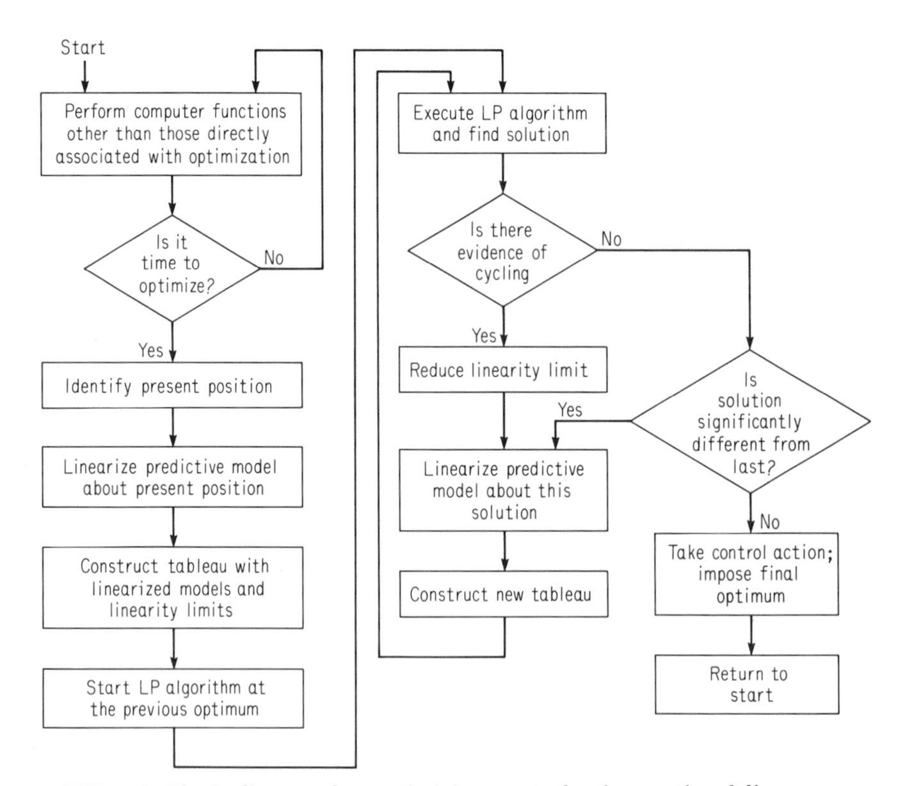

FIG. 3-4. Block diagram for optimizing control using sectional linear programming. Reproduced from E. S. Savas, Computer Control of Industrial Processes, © 1970 by McGraw-Hill, New York. Used with permission of the McGraw-Hill Book Company.

linear programming have very little in common besides both being optimization techniques. As developed by Bellman (28,29), dynamic programming is basically a technique for optimizing multistate decision processes. It has received a considerable amount of attention in the process industries (30), since most units consist of a sequence of stages. For optimization problems for such units, dynamic programming converts a high-dimensional optimization problem into a series of lower-dimensional optimization problems.

The computational difficulty with applying dynamic programming to practical problems is due to its rather large storage requirements, which Bellman has termed the "curse of dimensionality." Thus only a few applications have appeared (31). But if future generations of computer hardware live up to expectations, their larger core memories will enable this technique to be applied more often.

Lacking much of the mathematical elegance of linear or dynamic programming, *multivariable search techniques* are quite effective optimization techniques. While having some mathematical basis, many are nothing more than a systematic sequence of steps that will eventually locate the optimum. There is invariably some trial-and-error along the way, leading to some wasted effort.

For most practical purposes, this search for the optimum can be envisioned as a mountain-climbing scheme, the objective in both cases being to attain the peak. For a two-dimensional search, the criterion function surface can be represented in the same manner as elevation lines on a contour map. For example, Fig. 3-5 shows a response surface for a hypothetical reaction. The contour surfaces appear as slightly distorted elipses, which are not necessarily typical.

Such contour surfaces may be of great help in qualitatively evaluating the efficiency of search techniques. For example, a natural search strategy to propose is to use sequential univariable searches. That is, beginning with some given point, search in one direction until the optimum along this line has been found. Then repeat along the line in the other direction through the point resulting from the first search. This is again repeated until no further improvement is possible. Suppose this strategy is initiated at point A in Fig. 3-5. Searching first in the x-direction brings us to point B, the optimum along this line. Now searching in the y-direction fails to yield further improvement, thereby terminating the search. Obviously, the true optimum is not located, leading to the conclusion that this type of search is not very effective.

Criterion function surfaces such as in Fig. 3.5 are generally termed strongly unimodal, a term usually applied when the surface is unimodal along every line in the x-y plane. When this latter condi-

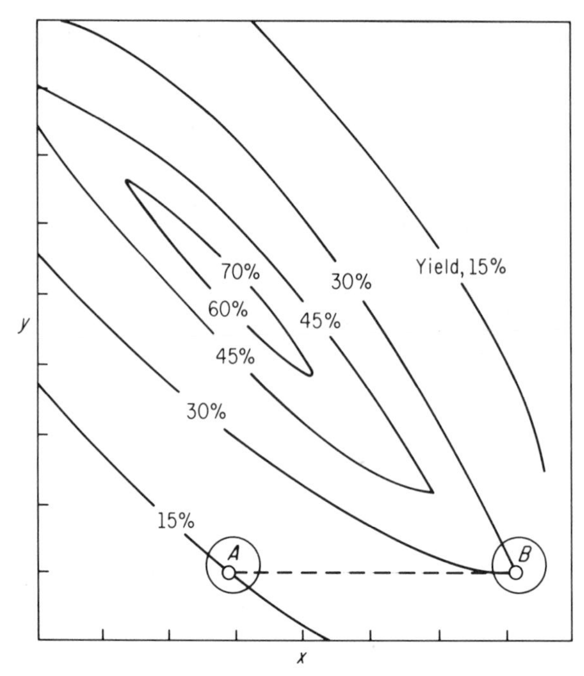

FIG. 3-5. Response surface for a chemical reaction. (Reproduced by permission from Ref. 2.)

tion is not true, it only suggests that the surface is not strongly unimodal, not that more than one optimum exists. For example, the criterion function surface in Fig. 3-6 is unimodal, but is not unimodal along the line AB.

All the search techniques to be discussed in this section are designed to locate only local optima. This is not to imply that multimodal functions are not encountered in everyday practice, but that there is simply no search technique that guarantees to locate the global optimum. One common approach for testing to see if local optima exist is to start the search from several different points. If they all terminate at the same point, local optima probably do not exist, this probability being directly proportional to the number of different points from which the search is started.

Search techniques can be classified into two categories: (1) gradient methods, and (2) pattern searches. The former rely on the gradients to determine the direction for the search, whereas the latter rely solely on excursions about the best point located so far in the search. Although we shall not consider any specific search techniques, we shall present the general characteristics of the methods in each category.

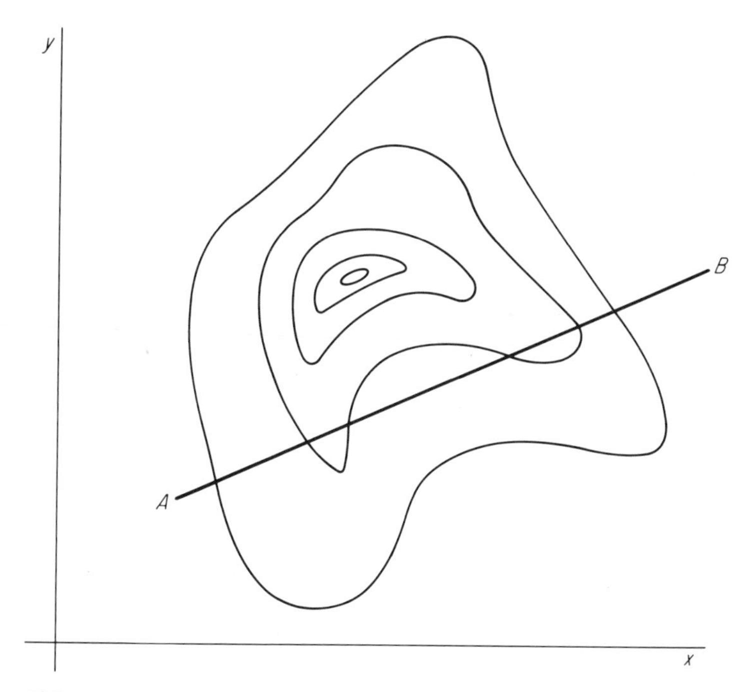

FIG. 3-6. Criterion function surface that is not strongly unimodal.

Gradient Methods. Perhaps the classical gradient method is that of steepest ascent from any given point. As evidenced by curve A in Fig. 3-7, this direction is always perpendicular to the contour lines. If continued long enough, this procedure will eventually arrive at a local maxima, at which the gradients are zero.

Although this method is simple in concept, it has some computational disadvantages. First, it tends to be laborious. As the gradients are normally calculated by finite differences, $n + 1$ evaluations (a base point plus n pertubations) are necessary to calculate the n partial directives that comprise the gradient at each point. This prohibits the recalculation of the gradient at too many points. From a given point, a univariable search is undertaken along the direction of the gradient until a maximum is approximately located. The gradient is reevaluated and the procedure repeated. This gives rise to curve B in Fig. 3-7.

Several techniques have been proposed for improving the gradient techniques. Some, e.g., PARTAN (parallel tangents), incorporate some pattern features into the search, while others, e.g., Davidson's, effectively approximate the second partials without actually evaluating them. Table 3-1 gives a list of various methods and the principal literature references.

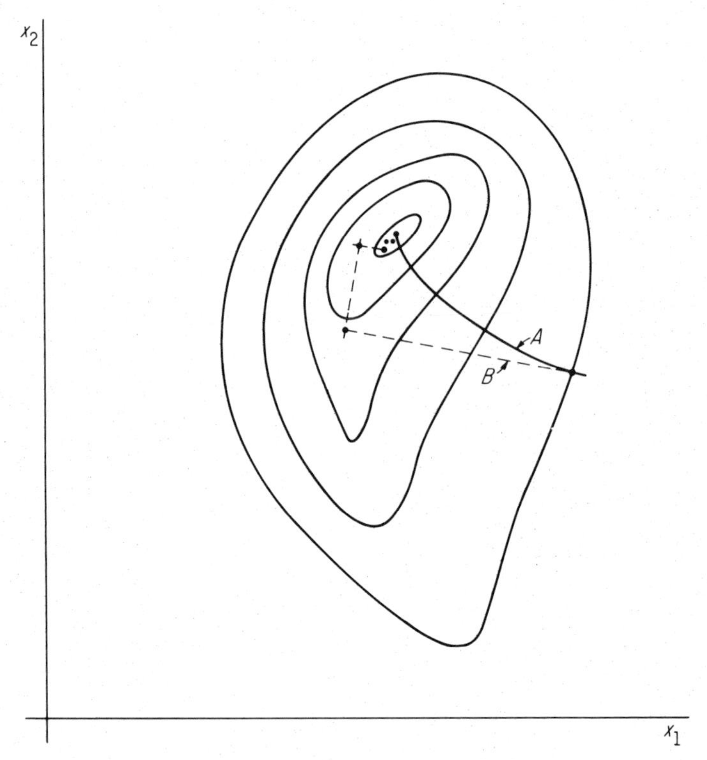

FIG. 3-7. Hill climbing via steepest ascent.

Pattern Searches. The general goal of these techniques is to make trial steps in some chosen directions to look for improvement. As the gradients are not calculated, they can sacrifice some incorrect steps and still potentially be as efficient as the gradient techniques. Searches of this type frequently have little mathematical basis, but have been found to be extremely successful. Depending upon the techniques, they may take their steps in predetermined directions (Pattern), in orthogonal directions modified as the search proceeds (Rosenbrock's), or in random directions (Random Search). The primary literature references are given in Table 3-1.

3-6 EVOLUTIONARY OPTIMIZATION TECHNIQUES

In the discussion of techniques presented previously in this chapter, it was tacitly assumed that a mathematical model of the process to be optimized was available. However, many processes need to be optimized that for various reasons do not lend themselves to mathe-

TABLE 3-1

Gradient Methods
1. Rotational Discrimination Law, V. J., and R. H. Fariss, "Process Optimization via Rotational Discrimination," paper 30c, 61st National A.I.Ch.E. Meeting, Houston, February 1967.
2. Davidon's Method Davidon, W. C., "Variable Metric Method for Minimization," A.E.C. Research and Development Report, ANL-5990, 1959.
3. Fletcher-Powell modification of Davidon's Method Fletcher, R., and M. J. D. Powell, "A Rapidly Convergent Descent Method for Minimization," *Computer Journal*, Vol. 6 (1963), p. 163.
4. Optimum Gradient Bekey, G. A., "Optimization of Multiparameter Systems by Hybrid Computer Techniques," Part II, *Simulation*, March 1964, p. 21.
5. Marquardt's Method Marquardt, D. W., "An Algorithm for Least-Squares Estimation of Nonlinear Parameters," *SIAM Journal*, Vol. 11, No. 2 (June 1963), p. 431.
6. Partan Shah, B. V., R. J. Buehler, and O. Kempthorne, "Some Algorithms for Minimizing a Function of Several Variables," *Journal of SIAM*, Vol. 12 (1964), p. 74.
7. Conjugate Gradient Fletcher, R. and C. M. Reeves, "Function Minimization by Conjugate Gradients," *Computer Journal*, Vol. 7 (1964), p. 149.
Pattern Methods
1. Coordinate Rotation Rosenbrock, H. H., "An Automatic Method for Finding the Greatest or Least Value of a Function," *Computer Journal*, Vol. 3 (1960), p. 175.
2. Pattern Search Wilde, D. J., *Optimum Seeking Methods*, Prentice-Hall, 1964.
3. Powell's Method Powell, M. J. D., "An Efficient Method for Finding the Minimum of a Function of Several Variables without Calculating Derivatives," *Computer Journal*, Vol. 7 (1965), p. 303.
4. Random Search Bekey, G. A., M. H. Gran, A. E. Sabroff, and A. Wong, "Parameter Optimization by Random Search Using Hybrid Computer Techniques," *Proceedings FJCC*, 1966, p. 191.

matical modeling. One example is the compressor station illustrated in Fig. 3-8. The objective is to regulate the speed of the drivers so that the desired discharge pressure is maintained with the minimum consumption of fuel.

If the two compressors had the same characteristics, the speeds of the two drivers should be equal at the optimum. If we require the

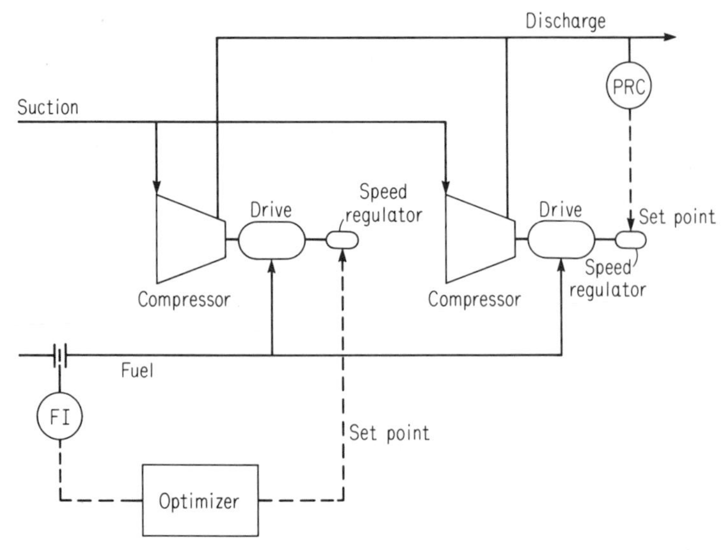

FIG. 3-8. Compressor station.

two speeds to be equal, there is only one value that will maintain the proper discharge pressure, and we no longer have an optimization problem. However, even if the two compressors are "identical," their characteristics will be slightly different and the true optimum will not occur when they are operated at the same speed.

Note that this is a constrained optimization problem, the constraint being the desired discharge pressure. If the characteristics of the two compressors were accurately known, the optimum could be calculated. However, the characteristics are rarely known to high accuracy, and even so they tend to change with time due to wear. It is in these kinds of problems that EVOP (evolutionary optimization) techniques are useful (22).

The basic idea of the EVOP approach is to use the process itself to evaluate the objective function. Note in Fig. 3-8 that the fuel flow, the quantity to be minimized, can be measured directly. Thus if the speeds of two compressors are set, the measured fuel flow is the cost function for this combination. However, the two speeds cannot be independently set in this case due to the constraint of maintaining the desired discharge pressure. But if one speed is set independently, there is one value of the speed of the other driver that will maintain the desired discharge pressure. As illustrated in Fig. 3-8, a conventional feedback controller can be used to regulate the speed of the second driver to maintain the desired discharge pressure.

The function of the optimizer in Fig. 3-8 is to determine the speed of the first driver that minimizes the fuel flow. Suppose the following strategy is used:

1. From an initial estimate of the proper speed, the value of the fuel flow is measured.
2. Using a predetermined step size, the value of the speed of the first drive is changed.
3. After enough time has elapsed for the feedback controller to adjust the speed of the second driver to maintain the desired discharge pressure, the fuel flow is again measured.
4. If the fuel flow is decreasing, then the optimizer is changing the speed of the first driver in the right direction and another step is made in the same direction. If the fuel flow is increasing, then the speed is varied in the opposite direction.

This is essentially a one-dimensional search problem being executed in real time on an operating process.

Although a bit more complicated, a similar approach can be devised for higher-dimensional problems, e.g., three or more compressors. In modifying conventional search techniques for this purpose, two factors must be considered, both arising from uncertainties in the measured value of the objective function.

1. As gradient techniques use small differences between values of the objective function evaluated at two slightly different operating points, they are susceptible to significant distortion due to disturbing influences.
2. Since some disturbances are large yet of short duration, a value of the objective function determined at some operating condition should not be considered valid over a long period of time. For example, if the value of the objective function increases for every possible perturbation about some operating point, the objective function should be reevaluated at the base operating period in case some disturbance has entered the process since the last measurement was made.

EVOP procedures have their advantages and disadvantages. For the long time constant processes encountered in the process industries, the evolutionary optimization process is time consuming. However, it can be permitted to run on a continuing basis to constantly adjust for disturbing factors, e.g., changing discharge flows in the compressor example presented earlier.

Although the EVOP procedures are a natural to implement digitally, at least one analog implementation is available (33).

LITERATURE CITED

1. Savas, E. S., *Computer Control of Industrial Process*, McGraw-Hill, New York, 1965.

2. Lee, T. H., G. E. Adams, and W. M. Gaines, *Computer Process Control: Modeling and Optimization*, Wiley, New York, 1968.
3. Cornish, H. L., et al., *Computerized Process Control*, Hobbs, Dorman & Company, New York, 1968.
4. Smith, C. L., R. W. Pike, and P. W. Murrill, *Development and Utilization of Mathematical Models*, Intext Educational Publishers, Scranton, Pa., 1970.
5. Franks, R. G. E., *Mathematical Modeling in Chemical Engineering*, Wiley, New York, 1968.
6. Mickley, H. S., T. K. Sherwood, and C. E. Reed, *Applied Mathematics in Chemical Engineering*, McGraw-Hill, New York, 1967.
7. Himmelblau, D. E., and K. B. Bischoff, *Process Analysis and Simulation: Deterministic Systems*, Wiley, New York, 1968.
8. McCabe, W. L., and J. C. Smith, *Unit Operations of Chemical Engineering*, second edition, McGraw-Hill, New York, 1967.
9. Levenspiel, O., *Chemical Reaction Engineering*, Wiley, New York, 1962.
10. Bird, R. B., W. E. Stewart, and E. N. Lightfoot, *Transport Phenomena*, Wiley, New York, 1960.
11. Roberts, S. M., and C. G. Laspe, "On-Line Computer Control of Thermal Cracking," *Industrial and Engineering Chemistry*, Vol. 53, No. 5 (May 1961), pp. 343–348.
12. Petryachuk, W. F., and A. I. Johnson, "The Simulation of An Existing Ethane-Dehydrogenation Reactor," *Canadian Journal of Chemical Engineering* Vol. 46, June 1960, pp. 172–181.
13. Shah, M. J., "Computer Control of Ethylene Production," *Industrial and Engineering Chemistry*, Vol. 59, No. 5 (May 1967), pp. 70–85.
14. Freedman, B., "Simulation of a Steam-Cracking, Gaseous Feed Ethylene Furnace," Proceedings of the Joint Automatic Control Conference, Atlanta, Georgia, June 1970.
15. Buchner, L. R., et al., "Modeling and Optimization of TCC for Computer Control," presented at the Midyear Meeting of the API's Division of Refining, pp. 337–354.
16. Shah, M. J., and R. E. Stillman, "Computer Control and Optimization of a Large Methanol Plant," *Industrial and Engineering Chemistry*, Vol. 62, No. 12 (December 1970), pp. 59–75.
17. Shah, M. J., "Control Simulation in Ammonia Production," *Industrial and Engineering Chemistry*, Vol. 59, No. 1 (January 1967), pp. 72–83.
18. Baxley, R. A., "Local Optimizing Control for Distillation," *Instrumentation Technology*, Vol. 16, No. 10 (October 1969), pp. 75–80.
19. Lupfer, D. E., and J. H. Engel, "Design of Automatic Control Systems to Operate Distillation Columns at Maximum Load," presented at the 68th National A.I.Ch.E. Meeting, Houston, Texas, Feb. 28–Mar. 4, 1971.
20. Rijnsdorp, J. E., and A. Maarleveld, "Optimizing Distillation by Constraint Control," presented at the IFAC Meeting, Warsaw, Poland, June 1969.
21. Wilde, D. J., *Optimum Seeking Methods*, Prentice-Hall, Englewood Cliffs, N.J., 1964.
22. Wilde, D. J., and C. S. Beightler, *Foundations of Optimization*, Prentice-Hall, Englewood Cliffs, N.J., 1967.

23. Thomas, M. E., and R. L. Gue, *Mathematical Methods in Operational Research*, Macmillan, New York, 1968.
24. Bay, T., C. W. Ross, J. C. Andrews, and J. L. Gilliland, "Breakthrough at Tijeras," *Leeds & Northrup Technical Journal*, Fall Issue, 1967, pp. 2-15.
25. Kirchanayer, L. K., *Economic Operation of Power Systems*, Wiley, New York, 1958.
26. Grabbe, E. M., S. Ramo, and D. E. Wooldridge, *Handbook of Automation Computation and Control*, Vol. 3, Wiley, New York, 1961.
27. Farmer, E. D., K. W. James, F. Moran, and P. Pettit, "Development of Automatic Digital Control of a Power System from the Laboratory to a Field Installation," *Proc. LEEE*, Vol. 116, March 1969, pp. 436-444.
28. Bellman, R. E., *Dynamic Programming*, Princeton U. P., Princeton, N.J., 1957.
29. Bellman, R. E., and S. E. Dreyfus, *Applied Dynamic Programming*, Princeton U. P., Princeton, N.J., 1962.
30. Roberts, S. M., *Dynamic Programming in Chemical Process Control*, Academic Press, New York, 1964.
31. Rees, F. J., and R. E. Larson, "Application of Dynamic Programming to the Optimal Dispatching of Electric Power from Multiple Types of Generation," *Proc. JACC*, Atlanta, Ga., June 1970, pp. 19-28.
32. Baasel, W. D., "Exploring Response Surfaces to Establish Optimum Conditions," *Chemical Engineering*, Oct. 25, 1965, pp. 147-152.
33. Turner, A. E., "New Horizons for Analog Control," presented at the 23rd Annual Symposium of the ISA Process Measurement Controls Division, Rutgers University, Mar. 23-25, 1970.

chapter **4**

Mathematics of Sampled-Data Systems

The objective of this chapter is to introduce the mathematical concepts applicable to discrete, digital, or sampled-data systems such as a direct digital control loop. The primary tool for the analysis of digital systems is the z-transform, which we shall see is very similar to and in many respects analogous to the Laplace transform.

Our approach to z-transforms is quite different from that used in most previous works (1–6). We shall, however, take a brief look at this other approach in Chapter 5.

4-1 THE DIGITAL CONTROL LOOP

Figure 4-1 illustrates the basic control loop encountered in direct digital control systems. Starting with the output of the sensor, we shall proceed around the loop, examining each element in turn.

The basic function of the input multiplexer is to obtain values of a function at certain intervals, an operation known as *sampling*. Although some samplers operate at irregular rates and others even randomly, the discussion in this chapter will be restricted to those that sample regularly with time T between successive entries. The majority of the industrial computer control loops are regular, with T in the range of 1 sec to 5 min, depending upon the process.

The output of the sensor is a continuous function of time, which we shall denote by $c(t)$. The output of the input multiplexer is a

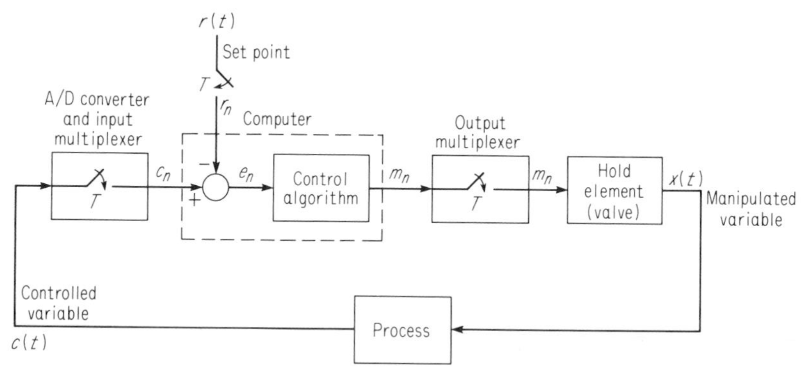

FIG. 4-1. Typical digital control loop.

number sequence which we shall represent as c_n. The subscript denotes the sampling instant, i.e., $c_1 = c(T)$, $c_2 = c(2T)$, etc. This is illustrated in Fig. 4-2.

The computer compares the output of the input multiplexer with a desired value or set point retrieved from some storage location within the computer itself. The sampling operation on the set point is thus accomplished by the computer's processing unit. In fact, the set point is rarely obtained by sampling a continuous signal as the diagram in Fig. 4.1 would indicate, but this representation is frequently used to emphasize that the number sequence c_n is compared to another number sequence r_n, the difference being the error $e_n = r_n - c_n$.

Using the appropriate control algorithm, the computer calculates a value for the manipulated variable from the error sequence e_n. Since this computation is done only at certain intervals of time, the output of the control algorithm is another number sequence which we shall refer to as m_n.

The output multiplexer transfers this value to the input of the hold element, whose function is to construct (from the number sequence m_n) a continuous signal $x(t)$ that becomes the input to the process. Since the process is typically continuous, it cannot accept values of m_n directly. The hold logic normally used is that of the zero-order hold, which maintains the last value of the output during the following sampling interval, i.e.,

$$x(t) = m_i, \qquad iT \leqslant t \leqslant (i+1)T$$

This is illustrated in Fig. 4-2. We shall consider this hold in more detail in a subsequent section.

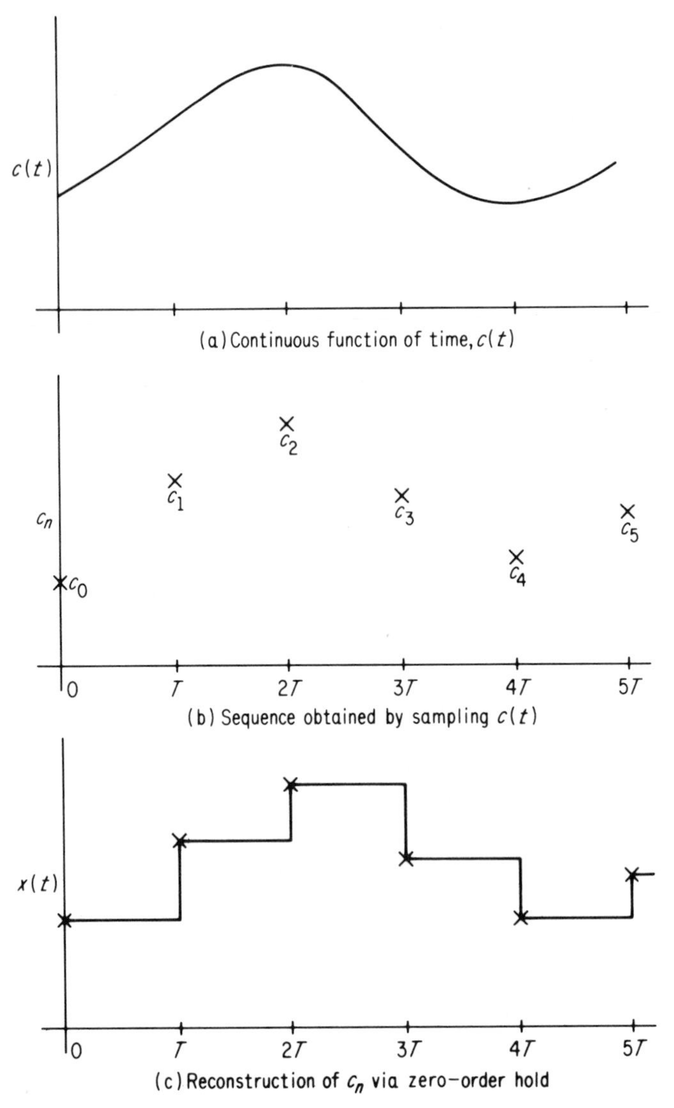

FIG. 4-2. Sampling and reconstruction process.

4-2 MATHEMATICAL ANALYSIS OF DIGITAL CONTROL LOOPS

The control loop in Fig. 4-1 contains two types of signals. From the output of the hold to the input to the input multiplexer, the signal is continuous. From the output of the input multiplexer to the input to the hold, the signal is discrete. For systems with continuous inputs and continuous outputs, differential equations describe their behavior. Systems with discrete inputs and discrete outputs are described by difference equations.

In the analysis of continuous systems such as the process, Laplace transforms have been used extensively in the control field, even though this approach is limited to linear systems. For example, the process in Fig. 4-1 can be represented as follows:

$x(t)$, $X(s)$ → [$g(t)$, $G(s)$] → $c(t)$, $C(s)$

This relationship is expressed mathematically via Laplace transforms as follows:

$$C(s) = G(s)X(s)$$

where

$$X(s) = \mathcal{L}\{x(t)\} = \int_0^\infty e^{-st}x(t)\,dt$$

$$G(s) = \mathcal{L}\{g(t)\} = \int_0^\infty e^{-st}g(t)\,dt$$

$$C(s) = \mathcal{L}\{c(t)\} = \int_0^\infty e^{-st}c(t)\,dt$$

Alternatively, this relationship can be expressed by the convolution integral

$$c(t) = \int_0^t g(\tau)x(t-\tau)\,d\tau$$

$$= \int_0^t x(\tau)g(t-\tau)\,d\tau$$

In addition, the output $c(t)$ can be related to $x(t)$ via a differential equation.

For discrete systems, analogous representations are available. For example the control algorithm in Fig. 4.1 can be represented as follows:

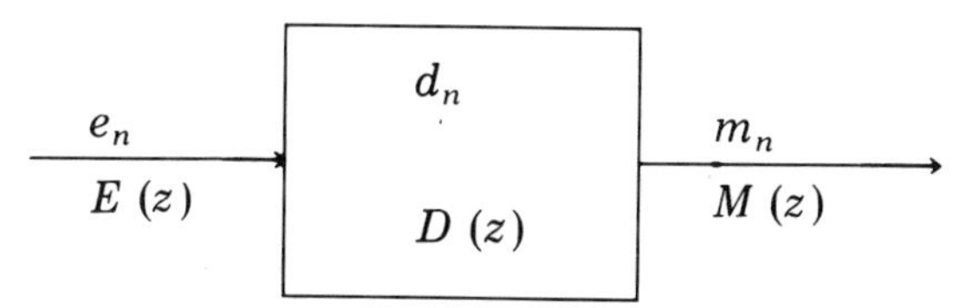

Analogous to the convolution integral representation for continuous systems, the discrete system can be represented as

$$m_n = \sum_{i=0}^{n} e_i d_{n-i} = \sum_{i=0}^{n} e_{n-i} d_i$$

Alternatively, this relationship can be expressed in terms of z-transforms.

$$M(z) = D(z)E(z)$$

where
$$E(z) = \mathfrak{z}\{e_n\}$$
$$D(z) = \mathfrak{z}\{d_n\}$$
$$M(z) = \mathfrak{z}\{m_n\}$$

We shall define the z-transform in the next section. In a later section, we shall also show how to express a difference equation in terms of z-transforms, and vice versa.

4-3 THE z-TRANSFORM

If f_n is a number sequence, its z-transform is defined as follows:

$$\begin{aligned} F(z) = \mathfrak{z}\{f_n\} &= \sum_{i=0}^{\infty} f_i z^{-i} \\ &= f_0 + f_1 z^{-1} + f_2 z^{-2} + f_3 z^{-3} + \cdots \\ &= f(0) + f(T)z^{-1} + f(2T)z^{-2} + f(3T)z^{-3} + \cdots \end{aligned} \tag{4-1}$$

The use of z-transforms for discrete signals is quite analogous to the use of Laplace transforms for continuous signals. Although the definition of the z-transforms is really based on the discrete signal f_n, it is customary to talk about the z-transform of continuous signals. However, since the definition of the z-transform is based on discrete signals, what is meant is the z-transform of the discrete signal obtained by sampling the continuous signal. For example, we will write $\mathfrak{z}\{e^{at}\}$ when we really mean $\mathfrak{z}\{e^{anT}\}$, where e^{anT} is the sequence obtained by sampling the signal e^{at} using a sampling time of T.

For the first example, suppose $f(t)$ is the unit step function $u(t)$. By definition,

$$F(z) = \mathfrak{z}[u(t)] = \sum_{n=0}^{\infty} u(nT)z^{-n} = \sum_{n=0}^{\infty} z^{-n} = 1 + z^{-1} + z^{-2} + \cdots$$

Recall that the sum of a geometric progression $\sum_{i=0}^{\infty} ar^i$, equals $a/(1 - r)$, where a is the first term and $r < 1$ is the ratio between successive terms. Thus

$$F(z) = \frac{1}{1 - z^{-1}}, \quad |z^{-1}| < 1 \tag{4-2}$$

This can be verified by reference to the table of transforms in Appendix A.

Next, consider the z-transform of the exponential function e^{-at}. Again by definition,

$$F(z) = \mathfrak{z}[e^{-at}] = \sum_{n=0}^{\infty} e^{-anT} z^{-n}$$

$$= \sum_{n=0}^{\infty} (e^{aT} z)^{-n}$$

$$= \frac{1}{1 - z^{-1} e^{-aT}}, \quad |z^{-1}| < e^{aT} \tag{4-3}$$

This can also be verified in Appendix A.

It should be noted in Appendix A that the Laplace transforms are also tabulated alongside the time function and the corresponding z-transforms. This is to allow direct conversion from the Laplace transform of a continuous function to the z-transform of that same function. In fact, it is common to speak of the z-transform of a Laplace transform—that is, $\mathfrak{z}[F(s)]$. What is really meant is the z-transform of the inverse Laplace transform.

In addition to the table of z-transforms in Appendix A, the following properties of the z-transform are worthy of note.

1. The z-transformation is linear. First consider the z-transform of the sum of two functions $f(t)$ and $g(t)$:

$$\mathfrak{z}[f(t) + g(t)] = \sum_{n=0}^{\infty} [f(nT) + g(nT)]\, z^{-n}$$

$$= \sum_{n=0}^{\infty} f(nT) z^{-n} + \sum_{n=0}^{\infty} g(nT) z^{-n}$$

$$= F(z) + G(z) \tag{4-4}$$

Consequently, the z-transform of a sum of two functions is the sum of their individual z-transforms.

Next consider the z-transform of a constant c times a function $f(t)$:

$$\mathfrak{z}\,[cf(t)] = \sum_{n=0}^{\infty} cf(nT)z^{-n}$$

$$= c \sum_{n=0}^{\infty} f(nT)z^{-n}$$

$$= cF(z) \tag{4-5}$$

Thus the z-transform of the product of a constant and a function equals the product of the constant and the z-transform of the function.

Together these properties establish the linearity of the z-transformation.

2. Delays. Suppose the function $f(t)$ is delayed k sampling periods, the value of $f(t - kT)$ being restricted to zero for $t < kT$. This is conveniently accomplished mathematically by defining the delays as $f(t - kT)u(t - kT)$. The z-transform of a delayed function can be obtained.

$$\mathfrak{z}\,[f(t - kT)u(t - kT)] = \sum_{n=0}^{\infty} f[(n - k)T]\,u[(n - k)T]\,z^{-n} \tag{4-6}$$

Substituting $m = n - k$,

$$\mathfrak{z}\,[f(t - kT)u(t - kT)] = \sum_{m=-k}^{\infty} f[mT]\,u[mT]\,z^{-(m+k)} \tag{4-7}$$

$$= z^{-k} \sum_{m=0}^{\infty} f(mT)z^{-m} = z^{-k}F(z) \tag{4-8}$$

Expressing this in terms of the discrete sequences,

$$\mathfrak{z}\,\{f_{n-k}\} = z^{-k}F(z) \tag{4.9}$$

This will be used extensively when we consider difference equations in the next section.

Section 4-9 considers the case when the dead time does not equal an integer number of sampling times.

4-4 z-TRANSFORMS OF DIFFERENCE EQUATIONS

The algorithms used by the computer to determine control actions are in reality difference equations. For example, the equation describing the continuous PI controller is

$$m(t) = K_c \left[e(t) + \frac{1}{T_i} \int e(t)\, dt \right] + M_R \qquad (4\text{-}10)$$

where K_c = controller gain
T_i = reset time
M_R = initial value of $m(t)$

Using rectangular integration for the integral gives the following two discrete equations, one for m_n and one for m_{n-1}.

$$m_n = K_c e_n + \frac{K_c T}{T_i} \sum_{k=1}^{n} e_k + M_R \qquad (4\text{-}11)$$

$$m_{n-1} = K_c e_{n-1} + \frac{K_c T}{T_i} \sum_{k=1}^{n-1} e_k + M_R$$

Subtracting these two equations,

$$m_n - m_{n-1} = K_c (e_n - e_{n-1}) + \frac{K_c T}{T_i} e_n \qquad (4\text{-}12)$$

As will be discussed later, this is the velocity form of the PI algorithm. This equation can be viewed in the following manner:

1. The output m_n is a sequence of pulses, not just a single value. The z-transform of this pulse sequence is $M(z)$.
2. Delaying m_n by one sampling time gives m_{n-1}, whose z-transform is $z^{-1} M(z)$.
3. Viewing the sequences e_n and e_{n-1} in an analogous manner gives the z-transforms $E(z)$ and $z^{-1} E(z)$ for e_n and e_{n-1} respectively.

Since the z-transform is a linear operation, the z-transform of the PI velocity algorithm of Eq. 4-12 can be obtained by taking the z-transform of each individual term

$$M(z) - z^{-1} M(z) = K_c [E(z) - z^{-1} E(z)] + \frac{K_c T}{T_i} E(z)$$

Solving for the transfer function,

$$\frac{M(z)}{E(z)} = D(z) = K_c \left[1 + \frac{T}{T_i} \frac{1}{1 - z^{-1}} \right] \qquad (4\text{-}13)$$

Recall that the transfer function of the continuous PI algorithm is

$$D(s) = K_c \left(1 + \frac{1}{T_i s} \right)$$

By comparison, we see that the integrator or $1/s$ term is equivalent to the term $(1 - z^{-1})$ in the denominator of $D(z)$. In the z-domain, the requirement that the controller exhibit no offset (i.e., contains the integral mode) translates into the requirement that the denominator of the control algorithm contain a factor $(1 - z^{-1})$ that does not cancel with a factor in the numerator.

By following this procedure in reverse, a transfer function in the z-domain can be expressed as a difference equation. Illustrating this with the algorithm in Eq. 4-13, first cross-multiply to obtain

$$(1 - z^{-1})M(z) = \left[K_c(1 - z^{-1}) + \frac{T}{T_i}\right] E(z)$$

or

$$M(z) - z^{-1} M(z) = K_c [E(z) - z^{-1} E(z)] + \frac{K_c T}{T_i} E(z)$$

This equation can be inverted term by term with the following relationships:

$$\begin{aligned} \mathfrak{z}^{-1}\{M(z)\} &= m_n \\ \mathfrak{z}^{-1}\{E(z)\} &= e_n \\ \mathfrak{z}^{-1}\{z^{-1} M(z)\} &= m_{n-1} \\ \mathfrak{z}^{-1}\{z^{-1} E(z)\} &= e_{n-1} \end{aligned}$$

The final result is

$$m_n - m_{n-1} = K_c(e_n - e_{n-1}) + \frac{K_c T}{T_i} e_n$$

This equation is identical to Eq. 4-12.

4-5 PULSE TRANSFER FUNCTIONS

The discussion in the previous section centered around the control algorithm, which is basically a discrete relationship between a discrete input and a discrete output. For this element of the loop, we had no difficulty deriving the equivalent z-transform expression.

Next consider the continuous elements of the loop, namely the hold and the process as shown in Fig. 4-3. Although the hold and process are continuous elements, the input to the hold is the discrete

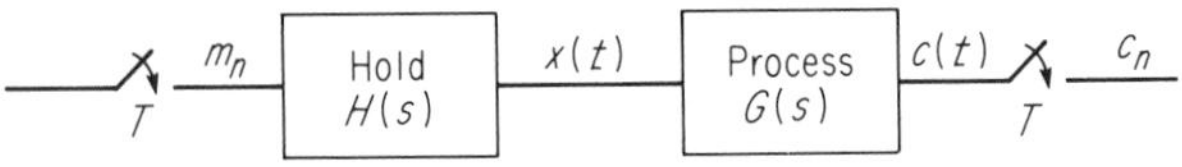

FIG. 4-3. Continuous elements of the DDC loop.

signal m_n and the output of the final sampler is the discrete signal c_n. Can these two sequences be related by a difference equation or by a z-transform expression? The answer is yes, the relationship in the z-domain being

$$\frac{C(z)}{M(z)} = HG(z) \tag{4-14}$$

where $HG(z)$ is called the *pulse transfer function.* We shall now relate the pulse transfer function to the continuous transfer functions $H(s)$ and $G(s)$.

First, consider the hold alone. In by far the majority of the process control applications, the zero-order hold is used. As stated previously, its performance is expressed by the relationship

$$x(t) = m_n, \qquad nT \leqslant t \leqslant (n+1)T$$

and was illustrated in Fig. 4-2.

Now consider deriving the transfer function $H(s)$ for the hold. Suppose we synthesize a signal p_n defined as follows:

$$p_n = \begin{cases} 1, & n = 0 \\ 0, & n \neq 0 \end{cases}$$

If this signal is the input to the hold, its output is illustrated in Fig. 4-4. Since the hold ideally moves instantaneously from zero to one at time zero, it is as if the unit impulse function $\delta(t)$ were imposed at that time. If no further inputs were made until time T, the

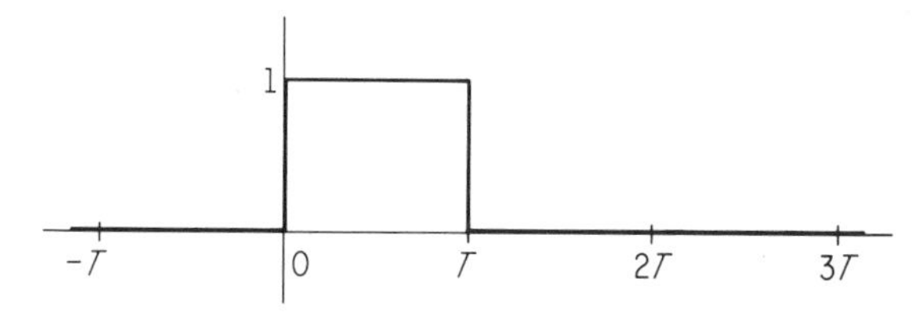

(a) Response of the zero-order hold to the input sequence p_n

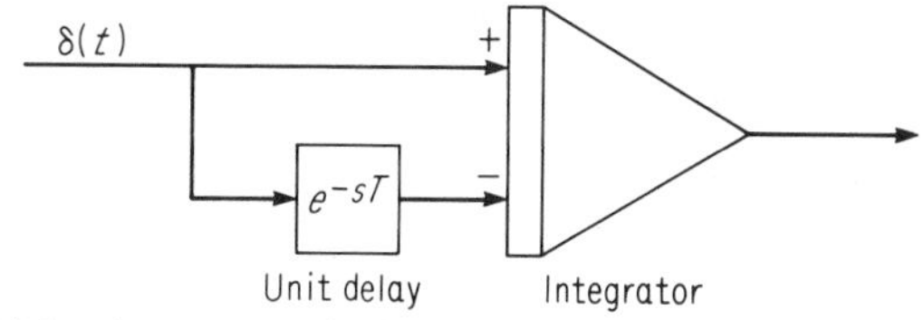

(b) Continuous network whose impulse response is identical to that of Fig. 4-3a

FIG. 4-4. Representation of the sampler and hold.

output of the integrator remains at one. If at that time, the input $-\delta(t)$ were made, the output would immediately return to zero and remain there. This procedure is illustrated in Fig. 4-4b.

Thus, subjecting the zero-order hold to the input p_n is equivalent to subjecting the circuit in Fig. 4-4b to the impulse function. The transfer function of the hold becomes that of the circuit in Fig. 4-4b, namely

$$H(s) = \frac{1 - e^{-sT}}{s} \tag{4-16}$$

and the response in Fig. 4-4a becomes its impulse response.

Extending this rationale to the process-hold combination in Fig. 4-3, the process response to the input p_n at the hold is given by

$$C(s) = H(s)G(s)$$

$$c(t) = \mathcal{L}^{-1}\{H(s)G(s)\}$$

The z-transform $C(z)$ is given by

$$C(z) = \mathfrak{z}\{c(t)\} = \mathfrak{z}\{\mathcal{L}^{-1}[H(s)G(s)]\}$$

which we also denote by

$$C(z) = \mathfrak{z}\{H(s)G(s)\}$$

The pulse transfer function is defined by

$$HG(z) = \frac{C(z)}{P(z)}$$

From the definition of the z-transform,

$$P(z) = p_0 + p_1 z^{-1} + p_2 z^{-2} + \cdots = 1$$

Therefore

$$HG(z) = C(z) = \mathfrak{z}\{H(s)G(s)\} \tag{4-17}$$

Thus we have related the pulse transfer function to the continuous elements of the loop.

Unfortunately, the relationship

$$\mathfrak{z}\{H(s)G(s)\} = H(z)G(z)$$

does *not* generally hold except for the special case when either $H(s)$ or $G(s)$ contains only delay terms.

As an example, suppose $G(s)$ is the simple first-order system $\dfrac{1}{s+1}$.

Then

$$HG(z) = \mathfrak{z}\left\{\frac{1 - e^{-sT}}{s} \cdot \frac{1}{s+1}\right\}$$

$$= \mathfrak{z}\left\{\frac{1}{s(s+1)}\right\} - \mathfrak{z}\left\{\frac{e^{-sT}}{s(s+1)}\right\}$$

Since the expression e^{-sT} in the second term is the unit delay it can be replaced by a coefficient z^{-1} as follows:

$$HG(z) = \mathfrak{z}\left\{\frac{1}{s(s+1)}\right\} - z^{-1}\mathfrak{z}\left\{\frac{1}{s(s+1)}\right\}$$

$$= (1 - z^{-1})\,\mathfrak{z}\left\{\frac{1}{s(s+1)}\right\}$$

Note that the same result is obtained if the term $1 - e^{-sT}$ in eq. 4-16 is replaced by the factor $1 - z^{-1}$. This should support the previous statement that the expression $\mathfrak{z}\{H(s)G(s)\} = H(z)G(z)$ is valid when either $H(s)$ or $G(s)$ contains only delay terms.

The next step is to expand the Laplace transform expression via the partial-fraction expansion:

$$HG(z) = (1 - z^{-1})\,\mathfrak{z}\left\{\frac{1}{s} - \frac{1}{s+1}\right\}$$

$$= (1 - z^{-1})\left[\mathfrak{z}\left\{\frac{1}{s}\right\} - \mathfrak{z}\left\{\frac{1}{s+1}\right\}\right]$$

Using the z-transform table in Appendix A to obtain expressions for $\mathfrak{z}\left\{\frac{1}{s}\right\}$ and $\mathfrak{z}\left\{\frac{1}{s+1}\right\}$ gives

$$HG(z) = (1 - z^{-1})\left[\frac{1}{1 - z^{-1}} - \frac{1}{1 - e^{-T}z^{-1}}\right]$$

$$= \frac{(1 - e^{-T})z^{-1}}{1 - e^{-T}z^{-1}}$$

This is the pulse transfer function.

For the system in Fig. 4-3,

$$HG(z) = \frac{C(z)}{M(z)} = \frac{(1 - e^{-T})z^{-1}}{1 - e^{-T}z^{-1}}$$

The equivalent difference equation can be obtained by cross-multiplying this expression to obtain

$$C(z) - e^{-T} z^{-1} C(z) = (1 - e^{-T}) z^{-1} M(z)$$

Recall that

$$c_n = \mathfrak{z}^{-1}\{C(z)\}$$
$$c_{n-1} = \mathfrak{z}^{-1}\{z^{-1} C(z)\}$$
$$m_{n-1} = \mathfrak{z}^{-1}\{z^{-1} M(z)\}$$

Using these to invert the previous equation to the time domain produces the difference equation

$$c_n - e^{-T} c_{n-1} = (1 - e^{-T}) m_{n-1}$$

This equation relates the output of the process at the sampling instants to the input to the hold. If only the outputs at the sampling instants are needed, the above difference equation will serve as well as the differential equation describing the process.

4-6 ANALYSIS OF BLOCK DIAGRAMS

Consider the typical block diagram of a sampled-data system as shown in Fig. 4-5. The analysis of this block diagram proceeds in a

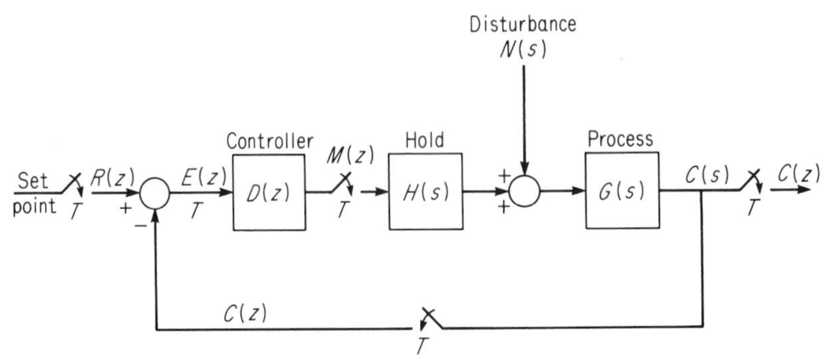

FIG. 4-5. Typical sampled-data block diagram.

manner analogous to the procedure for a completely continuous system. First, the sampled error $E(z)$ is given by

$$E(z) = R(z) - C(z)$$

The sampled output $C(z)$ is given by

$$\begin{aligned} C(z) &= \mathfrak{z}\{N(s)G(s)\} + D(z)\mathfrak{z}\{H(s)G(s)\}E(z) \\ &= NG(z) + HG(z)D(z)E(z) \end{aligned}$$

Substituting the first equation for $E(z)$ gives

$$C(z) = \frac{HG(z)D(z)R(z)}{1 + HG(z)D(z)} + \frac{NG(z)}{1 + HG(z)D(z)}$$

Note that the general form is very similar to that of the corresponding expression for a continuous system.

If the disturbance $N(s)$ is zero, this allows the previous equation to be solved for $C(z)/R(z)$, giving the closed-loop pulse transfer function $G_c(z)$

$$\frac{C(z)}{R(z)} = G_c(z) = \frac{HG(z)D(z)}{1 + HG(z)D(z)}$$

On the other hand, if $R(z)$ is zero, the resulting expression

$$C(z) = \frac{NG(z)}{1 + HG(z)D(z)}$$

cannot be solved for the ratio $C(z)/N(z)$, which is the pulse transfer function. Note that the difference is that the set point enters the loop through a sampler, whereas $N(s)$ does not. In these latter cases, a pulse transfer function as classically defined (output/input) cannot be obtained, unless $N(s)$ is restricted to be of a specified form, e.g., step, ramp, etc.

As an example, suppose the elements in Fig. 4-5 are defined as follows:

$D(z) = K$, a proportional controller

$$H(s) = \frac{1 - e^{-sT}}{s}$$

$G(s) = \dfrac{1}{s}$, a pure interator

$$N(s) = 0$$

$R(z) = \dfrac{1}{1 - z^{-1}}$, the unit step

The pulse transfer function is determined as follows:

$$\begin{aligned} HG(z) &= \mathfrak{z}\,[H(s)G(s)] = \mathfrak{z}\left[\frac{1 - e^{-sT}}{s^2}\right] \\ &= \mathfrak{z}\left[\frac{1}{s^2}\right] - z^{-1}\,\mathfrak{z}\left[\frac{1}{s^2}\right] \\ &= (1 - z^{-1})\,\mathfrak{z}\left[\frac{1}{s^2}\right] \end{aligned}$$

Note that the term $(1 - e^{-sT})$ can be simply taken outside the z-transformation, becoming $(1 - z^{-1})$. Using Appendix A to determine the z-transform of $1/s^2$ gives

$$HG(z) = (1 - z^{-1}) \left[\frac{Tz^{-1}}{(1 - z^{-1})^2} \right]$$

$$= \frac{Tz^{-1}}{1 - z^{-1}} = \frac{C(z)}{M(z)}$$

This can also be expressed as a difference equation by first cross-multiplying

$$C(z) - z^{-1} C(z) = Tz^{-1} M(z)$$

and then inverting.

$$c_n - c_{n-1} = Tm_{n-1}$$

This can be solved for c_n in terms of the previous input m_{n-1} and previous output c_{n-1}.

The closed-loop transfer function $C(z)/R(z)$ is given by:

$$\frac{C(z)}{R(z)} = \frac{HG(z)D(z)}{1 + HG(z)D(z)} = \frac{KTz^{-1}}{1 + (KT - 1)z^{-1}}$$

This equation can also be expressed in difference form.

$$c_n + (KT - 1)c_{n-1} = KTr_{n-1}$$

Again, the output c_n can be expressed in terms of the previous input r_{n-1} and previous output c_{n-1}.

4-7 INVERSE TRANSFORMS

In the preceding paragraphs the objective was to obtain the z-transform of a given time domain function. In this section the objective will be to obtain the time domain function corresponding to a given z-transform function. This latter manipulation is called the *inverse z-transform*, and is denoted by

$$f_n = \mathfrak{z}^{-1} [F(z)]$$

It should be noted that the inverse yields the sampled function f_n, not the continuous function $f(t)$. From the definition of $F(z)$, note that the only values of $f(t)$ involved are the values at the sampling instants, that is, f_n. Thus we should not expect any additional information about $f(t)$ other than the values at the sampling instant.

In other words, what is being said is that the inverse of $F(z)$

yields a unique function f_n. But as it is conceivable that the sampled function f_n could be derived from two different continuous functions $f_1(t)$ and $f_2(t)$, it follows that the inverse of $F(z)$ does not yield a unique continuous function $f(t)$.

Perhaps the simplest inverse to obtain is for functions $F(z)$ which appear in Appendix A. The inverse of these functions is simply the time domain function corresponding to the appropriate $F(z)$. However, most inverses encountered will not appear directly in Appendix A, but often some mathematical manipulations can be performed to obtain the tabulated functions. Fortunately, the inverse z-transformation is also a linear operation, i.e.,

$$\mathfrak{z}^{-1}[c_1 F(z) + c_2 G(z)] = c_1 f_n + c_2 g_n$$

One way to obtain the inverse of a function $F(z)$ is to express it by a partial fraction expansion such as:

$$F(z) = F_1(z) + F_2(z) + F_3(z) + \cdots$$

This is similar to the procedure for Laplace transforms, and the inverses of $F_1(z)$, $F_2(z)$, etc., are obtained from Appendix A. It should be noted that the inverse transformation gives the resulting time domain function only at the sampling instants.

Not all functions can be expressed in terms of partial fraction expansions whose terms appear in Appendix A. Even for functions that appear, the following technique is often preferable. The function $F(z)$ is typically a ratio of polynomials of the type

$$F(z) = \frac{a_0 + a_1 z + a_2 z^2 + \cdots + a_m z^m}{b_0 + b_1 z + b_2 z^2 + \cdots + b_n z^n}$$

If the denominator of this expression is divided into the numerator, the result is

$$F(z) = c_0 + c_1 z^{-1} + c_2 z^{-2} + \cdots$$

Some of the coefficients may be zero, but this is of no consequence. Recall that the definition of the z-transform is

$$F(z) = f(0) + f(T)z^{-1} + f(2T)z^{-2} + \cdots$$

Comparing these two expressions indicates $f(0) = c_0$, $f(T) = c_1$, etc. Thus the value of the inverse transform at the end of each sampling period can be obtained by division of numerator by denominator.

For functions $F(z)$ for which neither of the above are applicable, the complex integral formula can be applied (4,5). This technique is applicable to any function $F(z)$, but it will not be discussed here as it is seldom required for process control work. Other difficulties, e.g., irrational powers of z, can also usually be surmounted without too much difficulty.

As an example of the inverse transformation, consider the following function:

$$F(z) = \frac{3}{(1 - z^{-1})^2 (1 - 0.5z^{-1})}$$

Let the sampling period be unity. By partial-fraction expansion,

$$F(z) = \frac{6z^{-1}}{(1 - z^{-1})^2} + \frac{3}{1 - 0.5z^{-1}}$$

Obtaining the inverses of each of the individual functions from Appendix A yields

$$f(t) = 6t + 3e^{-0.694t}$$

where the constant in the exponential term is obtained from $e^{-0.694} = 0.5$.

As an alternate approach, $F(z)$ could have been rearranged:

$$F(z) = \frac{6}{2 - 5z^{-1} + 4z^{-2} - z^{-3}}$$

Dividing numerator by denominator via long division,

$$F(z) = 3 + 7.5z^{-1} + 12.75z^{-2} + 18.325z^{-3} + 24.1875z^{-4} + \cdots$$

Thus the values of $f(t)$ at the sampling times are

$$\begin{aligned} f(0) &= 3 \\ f(1) &= 7.5 \\ f(2) &= 12.75 \\ f(3) &= 18.325 \\ f(4) &= 24.1875, \text{ etc.} \end{aligned}$$

These same values are obtained when the analytic function is evaluated at the appropriate values of time.

4-8 STABILITY

In the Laplace domain, it was concluded that a system was stable if all poles were in the left-half s-plane except for possibly one at the origin. Suppose we consider the situation in the z-domain.

As for continuous transfer functions, discrete transfer functions can be expanded in terms of factors of the denominator. That is,

$$G(z) = \frac{A(z)}{B(z)} = \frac{A(z)}{1 + b_1 z^{-1} + b_2 z^{-2} + \cdots + b_n z^{-n}}$$

$$= \frac{A(z)}{(1 - r_1 z^{-1})(1 - r_2 z^{-1}) \cdots (1 - 4_n z^{-n})}$$

$$= \frac{C_1}{1 - r_1 z^{-1}} + \frac{C_2}{1 - r_2 z^{-1}} + \cdots + \frac{C_n}{1 - r_n z^{-n}}$$

assuming all roots are distinct. As for Laplace expressions, complex roots always occur in conjugate pairs. Any repeated roots can be treated in the same manner as for expressions in the Laplace domain.

To illustrate the influence of the location of the root on stability, consider the six roots illustrated in Fig. 4-6. Suppose the individual

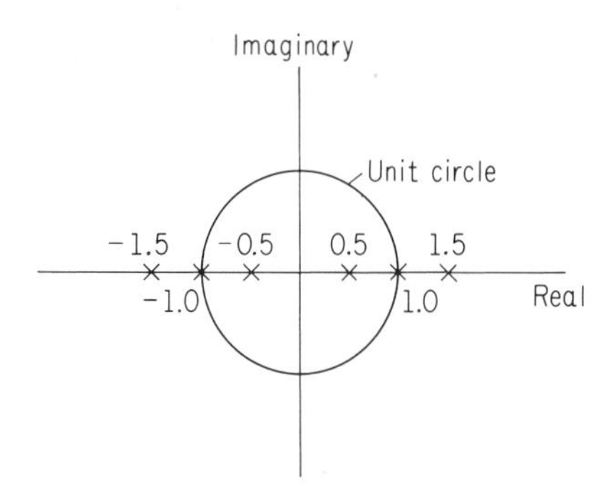

FIG. 4-6. The unit circle.

element corresponding to each of these roots is subjected to the sequence p_n defined earlier as

$$p_n = \begin{cases} 1 & n = 0 \\ 0 & n \neq 0 \end{cases}$$

First, consider the root located at $z = 0.5$, which corresponds to the term

$$\frac{1}{1 - 0.5z^{-1}}$$

Subjecting to the input p_n gives the system illustrated below:

$$p_n \longrightarrow \fbox{$\frac{1}{1 - 0.5z^{-1}}$} \longrightarrow g_n$$

The transfer function representation is

$$\frac{Y(z)}{P(z)} = \frac{1}{1 - 0.5z^{-1}}$$

Cross-multiplying and inverting yields the difference equation

$$Y(z) - 0.5z^{-1}\, Y(z) = P(z)$$

$$y_n - 0.5y_{n-1} = p_n$$

Starting with y_{-1} equal zero, this difference equation can be used to construct the following table:

n	p_n	y_n
0	1	1
1	0	0.5
2	0	0.25
3	0	0.125
4	0	0.0625
5	0	0.03125

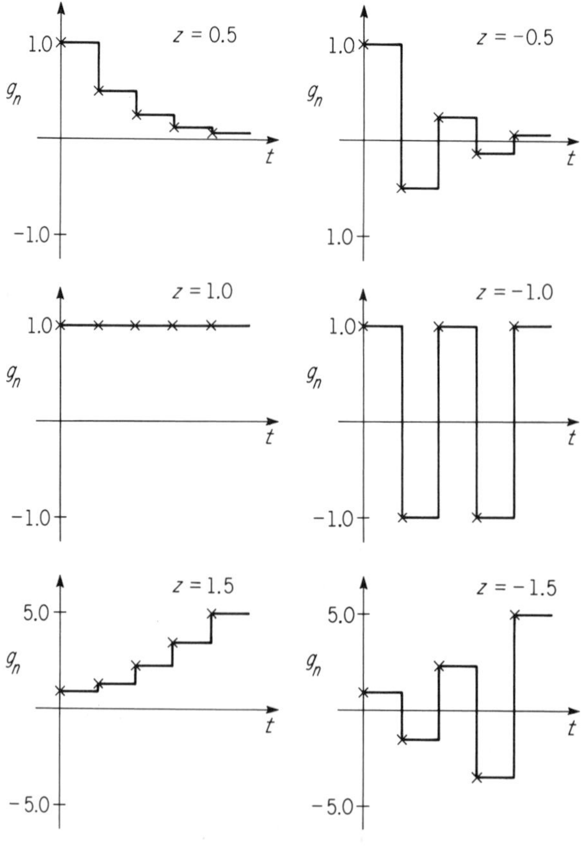

FIG. 4-7. Characteristic response of poles located at various points along the real axis (zero-order hold reconstruction added for clarity only).

This response is plotted in Fig. 4-7 along with the response of the elements corresponding to the other poles in Fig. 4-6.

From these plots, two factors become evident. First, any roots located more than one unit away from the origin lead to unstable responses. Although we have only illustrated this for real roots, it also applies to complex roots (see exercises). Thus the requirement for stability is that all roots lie within the unit circle except for possibly one at $z = 1$.

Second, roots lying on the negative real axis give responses for which successive values alternate in sign, producing a phenomenon known as *ringing*. In Chapter 6 we shall see that terms corresponding to these roots must be eliminated from control algorithms to obtain satisfactory performance.

4-9 SYSTEMS WITH DEAD TIME

One factor that distinguishes process systems from other types of systems is that a dead time (time delay or transportation lag) is very common in process systems. In fact, techniques that cannot be applied to systems with dead time are of marginal value in the analysis of process systems. For sampled-data systems, an extension of the z-transform known as the *modified z-transform* is directly applicable to systems with dead time.

Suppose our process is described by the following transfer function:

$$G_P(s) = G(s)\, e^{-\theta ' s}$$

where $G(s)$ contains no dead time. Furthermore, let

$$\theta = NT + \theta$$

where N is the largest integer number of sampling times in $\theta '$. Therefore,

$$0 < \theta < T$$

Taking the z-transform of $G_p(s)$,

$$G_p(z) = \mathfrak{z}[G(s)\, e^{-\theta ' s}] = z^{-N}\, \mathfrak{z}\, [G(s)\, e^{-\theta s}]$$

The z-transform of $G(s)e^{-\theta s}$ is given by the modified z-transform:

$$\mathfrak{z}\, [G(s)\, e^{-\theta s}] = \mathfrak{z}_m\, [G(s)] = G(z, m)$$

where $m = 1 - \theta / T$. In taking the modified z-transform, the dead time is implied to exist, and therefore we write $\mathfrak{z}_m\, [G(s)]$ instead of $\mathfrak{z}_m\, [G(s)e^{-\theta s}]$.

As an example, consider taking the modified z-transform of e^{-at}:

$$\mathfrak{z}_m[e^{-at}] = \mathfrak{z}_m\left[\frac{1}{s+a}\right] = \mathfrak{z}\left[\frac{e^{-\theta s}}{s+a}\right]$$

$$= \mathfrak{z}[e^{-a(t-\theta)}\, u(t-\theta)]$$

$$= \sum_{n=0}^{\infty} e^{-a(nT-\theta)}\, u(nT-\theta) z^{-n}$$

$$= e^{-a(T-\theta)} z^{-1} + e^{-a(2T-\theta)} z^{-2} + e^{-a(3T-\theta)} z^{-3} + \cdots$$

$$= e^{-amT} z^{-1} + e^{-amT} e^{-aT} z^{-2} + e^{-amT} e^{-2aT} z^{-3} + \cdots$$

$$= \frac{e^{-amT} z^{-1}}{1 - e^{-aT} z^{-1}}$$

This procedure can be applied to a series of functions to give the fourth column in the table in Appendix A.

However, one must be careful when using tables of modified z-transforms. As originally proposed, the modified z-transform was to provide a mechanism by which the response between sampling times could be calculated, an application that we shall illustrate shortly. For this purpose, either a delay (transfer function $= e^{-\theta s}$) or an advance (transfer function $= e^{\theta s}$) could be used, and modified z-transforms have been developed for both. But when the process actually contains dead time, the modified z-transform developed for the delay $e^{-\theta s}$ must be used rather than the one for the advance $e^{\theta s}$. The only modified z-transforms used in this text are those for the delay.

To illustrate the use of the modified z-transform, consider determining $C(z)/R(z)$ for the block diagram in Fig. 4-8. Following the

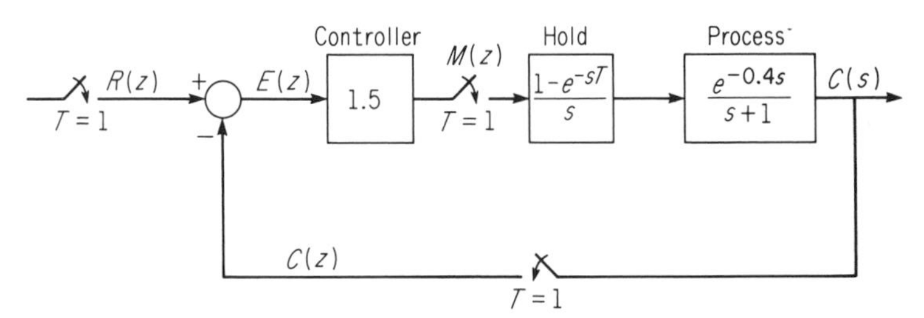

FIG. 4-8. Control loop with dead time.

procedure used in Sec. 4-6, we obtain

$$\frac{C(z)}{R(z)} = \frac{1.5HG(z)}{1 + 1.5HG(z)}$$

where

$$HG(z) = \mathfrak{z}\left[\frac{1 - e^{-sT}}{s} \cdot \frac{e^{-0.4s}}{s+1}\right]$$

$$= (1 - z^{-1})\,\mathfrak{z}\left[\frac{e^{-0.4s}}{s(s+1)}\right]$$

$$= (1 - z^{-1})\,\mathfrak{z}_m\left[\frac{1}{s(s+1)}\right]$$

where $m = 1 - \dfrac{\theta}{T} = 0.6$. Using the table of modified z-transforms in Appendix A,

$$HG(z) = (1 - z^{-1})\left[\frac{z^{-1}}{1 - z^{-1}} - \frac{e^{-mT}\,z^{-1}}{1 - e^{-T}\,z^{-1}}\right]$$

$$= z^{-1}\left[\frac{(1 - e^{-mT}) + (e^{-mT} - e^{-T})\,z^{-1}}{1 - e^{-T}\,z^{-1}}\right]$$

$$= \frac{z^{-1}\,(0.450 + 0.182z^{-1})}{1 - 0.368z^{-1}} = \frac{C(z)}{M(z)}$$

In difference equation form, this becomes

$$c_n - 0.368c_{n-1} = 0.450m_{n-1} + 0.182m_{n-2}$$

Thus we can express the output at the nth sampling instant in terms of the outputs and inputs at previous sampling instants.

The final expression for $C(z)/R(z)$ is

$$\frac{C(z)}{R(z)} = \frac{1.5z^{-1}\,(0.450 + 0.182z^{-1})}{1 - 0.368z^{-1} + 1.5z^{-1}\,(0.450 + 0.182z^{-1})}$$

$$= \frac{z^{-1}\,(1.125 + 0.455z^{-1})}{1 + 0.757z^{-1} + 0.455z^{-2}}$$

This is the transfer function for the closed-loop block diagram in Fig. 4-8.

4-10 OUTPUT BETWEEN SAMPLING INSTANTS

To use the modified z-transform to determine the output of the system, the block diagram in Fig. 4-9 is used. We would like to determine the response $c(t)$ for all values of t. Since the dead-time element in Fig. 4-9 is outside the loop, the process response $c(t)$ is inde-

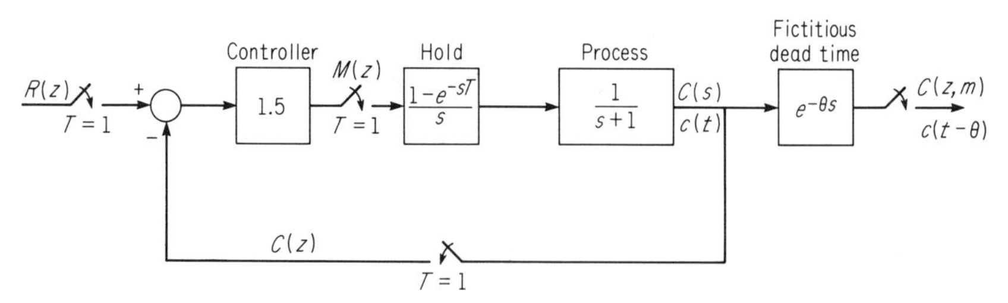

FIG. 4-9. Block-diagram representation used to determine outputs between sampling times.

pendent of the dead time. Its only function is to delay $c(t)$ so that we can "see" the intersample information.

To see how this is accomplished, consider the signals shown in Fig. 4-10. The actual continuous output of the system is shown in Fig. 4-10a. Only the values at the sampling instants would be obtained by the ordinary z-transform, as illustrated in Fig. 4-10a. However, if this signal is delayed before sampling, a different value is obtained, as illustrated in Fig. 4-10b. Advancing the output of the sampler gives a value between the sampled values of the original signal, as illustrated in Fig. 4-10c. As θ is varied from 0 to T, the entire signal between the sampling instants is obtained. Alternatively, m can be varied from 0 to 1 to obtain the intersample output.

To illustrate this procedure, the block diagram in Fig. 4-9 is represented by the following equation:

$$\frac{C(z,m)}{R(z)} = \frac{1.5HG(z,m)}{1 + 1.5HG(z)}$$

where

$$HG(z,m) = \mathfrak{z}_m\,[H(s)G(s)] = \mathfrak{z}\,[H(s)G(s)e^{-\theta s}]$$

$$= \frac{z^{-1}\,[(1 - e^{-mT}) + (e^{-mT} - e^{-T})z^{-1}]}{1 - e^{-T}z^{-1}}$$

$$HG(z) = \mathfrak{z}\,[H(s)G(s)]$$

$$= \mathfrak{z}\left[\frac{1 - e^{-sT}}{s(s+1)}\right]$$

$$= (1 - z^{-1})\,\mathfrak{z}\left[\frac{1}{s} - \frac{1}{s+1}\right]$$

$$= \frac{(1 - e^{-T})z^{-1}}{1 - e^{-T}z^{-1}}$$

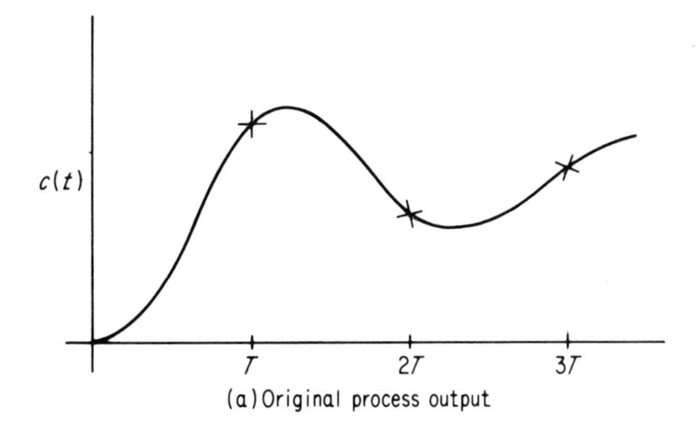

(a) Original process output

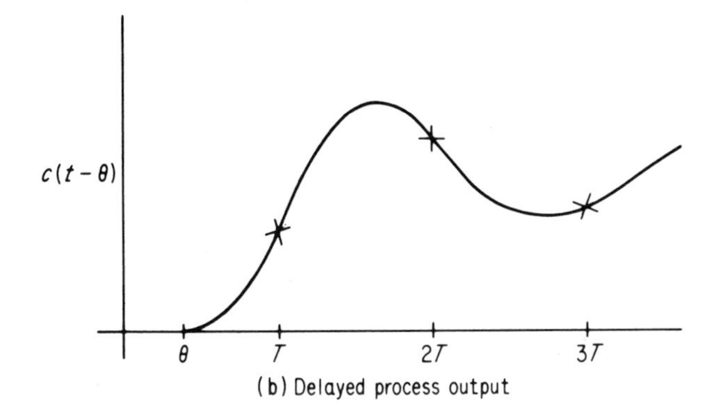

(b) Delayed process output

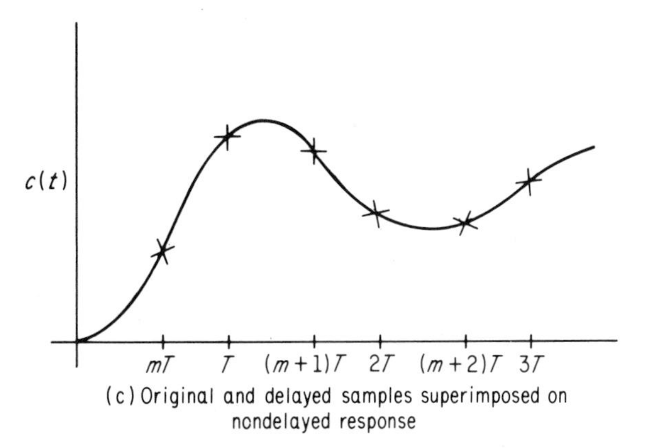

(c) Original and delayed samples superimposed on nondelayed response

FIG. 4-10. Concept behind using modified z-transform to determine values between sampling instants.

Therefore

$$\frac{C(z,m)}{R(z)} = \frac{1.5z^{-1}[(1-e^{-mT}) + (e^{-mT}-e^{-T})z^{-1}]}{1-e^{-T}z^{-1} + 1.5(1-e^{-T})z^{-1}}$$

$$= \frac{z^{-1}[(1-e^{-mT}) + (e^{-mT}-e^{-T})z^{-1}]}{1+0.58z^{-1}}$$

If $R(z)$ is the unit step, then

$$C(z,m) = \frac{1.5z^{-1}[(1-e^{-mT}) + (e^{-mT}-e^{-T})z^{-1}]}{[1+0.58z^{-1}][1-z^{-1}]}$$

$$= \frac{1.5z^{-1}[(1-e^{-mT}) + (e^{-mT}-e^{-T})z^{-1}]}{1-0.42z^{-1}-0.58z^{-1}}$$

By long division,

$$C(z,m) = 1.5(1-e^{-mT})z^{-1} + 1.5(0.58e^{-mT}+0.052)z^{-2} + 1.5(0.602-0.341e^{-mT})z^{-3} + 1.5(0.193e^{-mT}+0.383)z^{-4} + \dots$$

The result is illustrated in Fig. 4-11.

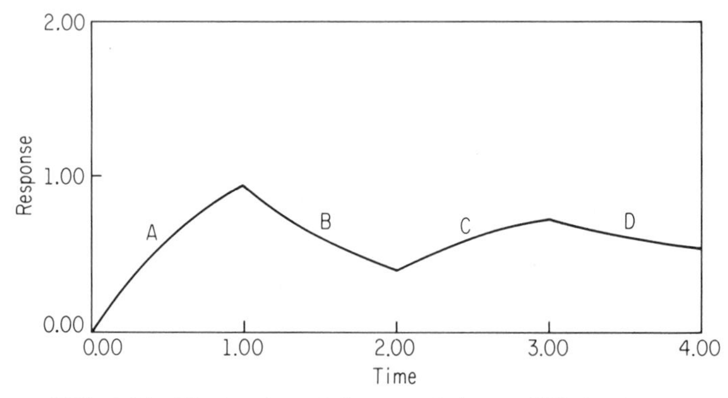

FIG. 4-11. Illustration of the use of the modified z-transform to calculate the output between sampling instants.

LITERATURE CITED

1. Lindorff, D. P., *Theory of Sampled-Data Control Systems*, Wiley, New York, 1965.
2. Jury, E. I., *Theory and Application of the z-Transform Method*, Wiley, New York, 1964.
3. Razazzini, J. R., and G. F. Franklin, *Sampled-Data Control Systems*, McGraw-Hill, New York, 1958.
4. Tou, J. T., *Digital and Sampled-Data Control Systems*, McGraw-Hill, New York, 1959.
5. Kuo, B. C., *Analysis and Synthesis of Sampled-Data Control Systems*, Prentice-Hall, Englewood Cliffs, N.J., 1963.
6. Monroe, A. J., *Digital Processes for Sampled-Data Systems*, Wiley, New York, 1962.

chapter 5

Frequency-Domain Considerations

The objective of this chapter is to bring together in one place in the text most of the important frequency considerations in the design and implementation of digital or sampled-data control systems. The first section discusses the effect of sampling on a function's frequency spectrum. The next section treats the frequency characteristics of holds as related to data reconstruction. The s-plane characteristics of the transform of a sampled function are then developed and extended to obtain a stability criterion in the z-domain. Characteristics of noise are developed and noise rejection procedures are related to them.

5-1 FREQUENCY SPECTRUM OF A SAMPLED SIGNAL (1)

A mathematical representation for a sampler can be readily accomplished using the impulse or delta function, $\delta(t)$, which is the derivative of the step function $u(t)$ shown in Fig. 5-1a. That is, $\delta(t) = du(t)/dt$. Thus its value is infinite at $t = 0$ but is zero elsewhere. If $\delta(t)$ is to be integrated to give the unit step function, the area of the impulse must be unity. If step functions of different heights are differentiated, the impulse will always be infinite in height but will have a different area. Thus the true measure of the impulse is its area. To graphically represent impulses, arrows are drawn whose height is proportional to the area of the impulse. The impulse function generated by differentiating the step function in Fig. 5.1a is shown in Fig. 5.1b.

To represent a sampler mathematically, it is necessary to represent a train of unit impulses (impulses whose area is unity) as shown

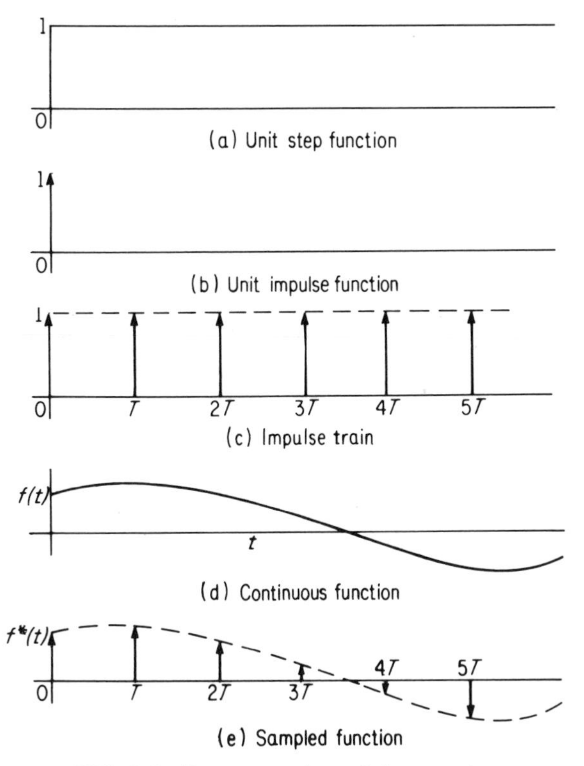

FIG. 5-1. Representation of the sampler.

in Fig. 5.1c. This train is most conveniently expressed by

$$\sum_{n=-\infty}^{\infty} \delta(t - nT)$$

where T is the period between samples. The sampled function is then represented by the product of the continuous function and the pulse train. The sampled function $f^*(t)$ obtained by sampling the continuous function $f(t)$ in Fig. 5.1d is shown in Fig. 5.1e. The area of each impulse in Fig. 5.1e equals the value of the continuous function at the sampling instant. Mathematically this is expressed

$$f^*(t) = f(t) \sum_{n=-\infty}^{\infty} \delta(t - nT)$$

$$= \sum_{n=-\infty}^{\infty} f(nT)\, \delta(t - nT)$$

Note that this function is zero except at the instants of sampling.

To develop the z-transform, the Laplace transform of $f^*(t)$ is taken.

$$\mathcal{L}\,[f^*(t)] = F^*(s) = \int_0^{\infty} \left[\sum_{n=-\infty}^{\infty} f(nT)\,\delta(t - nT) \right] e^{-st}\,dt$$

$$= \sum_{n=0}^{\infty} f(nT) e^{-nTs} \tag{5-1}$$

Introducing the change of variable $z = e^{sT}$ and defining the result as the z-transform $F(z)$ yields

$$F(z) = F^*(s)\,\big|_{z=e^{Ts}} = \sum_{n=0}^{\infty} f(nT) z^{-n}$$

This corresponds to the definition given in Eq. 4-1.

Letting $s = j\omega$ in the Laplace transform expression for the sampled function (Eq. 5.1) gives

$$F^*(j\omega) = \sum_{n=0}^{\infty} f(nT) e^{-nj\omega T} \tag{5-2}$$

Using this function, the Bode plot can be prepared for the sampled signal in a manner completely analogous to that for a continuous signal.

If the sampling time is T, the sampling frequency is $\omega_s = 2\pi/T$. Suppose we examine the frequency content at $j(\omega + m\omega_s)$, that is, at an integer number of sampling frequencies from ω. By Eq. 5-2,

$$F^*(j\omega + jm\omega_s) = \sum_{n=0}^{\infty} f(nT) e^{-n(j\omega T + jm\omega_s T)}$$

$$= \sum_{n=0}^{\infty} f(nT) e^{-nj\omega T}\, e^{-njm\omega_s T} \tag{5-3}$$

Examine the term $e^{-njm\omega_s T}$. As $\omega_s T = 2\pi$, this term becomes $e^{-j(nm\cdot 2\pi)}$. Recall that $re^{j\theta}$ is the representation of a point in the complex plane a distance r from the origin and at angle θ. Since the exponent $nm\cdot 2\pi$ of the term in question simply denotes an integer number of revolutions about the origin, the angle is zero. Thus, this term becomes unity at an angle of $0°$, or simply the real number one.

Substituting into Eq. 5-3 gives

$$F^*(j\omega + jm\omega_s) = \sum_{n=0}^{\infty} f(nT)\, e^{-nj\omega T}$$

or

$$F^*(j\omega + jm\omega_s) = F^*(j\omega)$$

This equation indicates that the frequency spectrum is a periodic function that is repeated over every interval of length ω_s along the frequency axis.

Next, consider expressing $F^*(j\omega)$ in terms of $F(j\omega)$. As discussed above, the sampling process can be mathematically represented as follows:

$$f^*(t) = p(t)\, f(t)$$

where

$$p(t) = \sum_{n=-\infty}^{\infty} \delta(t - nT)$$

Since $p(t)$ is a periodic function with period T, it can be represented by the exponential Fourier series:

$$p(t) = \sum_{n=-\infty}^{\infty} C_n e^{jn\omega_s T}$$

where C_n is the Fourier coefficient given by

$$C_n = \frac{1}{\mathrm{T}} \int_0^{\mathrm{T}} p(t) e^{-jn\omega_s t} dt$$

As the only component of $p(t)$ occuring over the interval from 0 to T is the impulse at zero, this equation becomes

$$C_n = \frac{1}{\mathrm{T}} \int_0^{\mathrm{T}} \delta(t) e^{-jn\omega_s t} dt = \frac{1}{T}$$

Thus using the exponential Fourier series for $p(t)$ gives

$$f^*(t) = \frac{1}{T} \sum_{n=-\infty}^{\infty} e^{jn\omega_s t} f(t)$$

Taking the Fourier transform of $e^*(t)$ gives

$$F^*(j\omega) = \mathfrak{F}[e^*(t)]$$

$$= \frac{1}{T} \sum_{n=-\infty}^{\infty} \mathfrak{F}[e^{jn\omega_s t} f(t)]$$

$$= \frac{1}{T} \sum_{n=0}^{\infty} F(j\omega + jn\omega_s)$$

Thus the frequency spectrum of the sampled function can be calculated from the frequency spectrum of the continuous function.

To illustrate this, consider $f(t)$ to be the bandlimited† signal whose amplitude spectrum is shown in Fig. 5.2a. Note that this figure is drawn for the case in which $\omega_c < \omega_s/2$. Sampling $f(t)$ at frequency ω_s gives the signal $f^*(t)$ whose frequency spectrum is shown in Fig. 5.2b. Note that the frequency spectrum of the con-

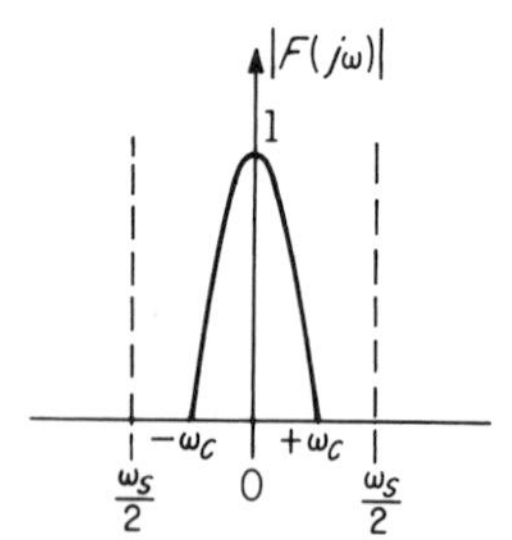

(a) Amplitude spectrum of the continuous function $f(t)$.

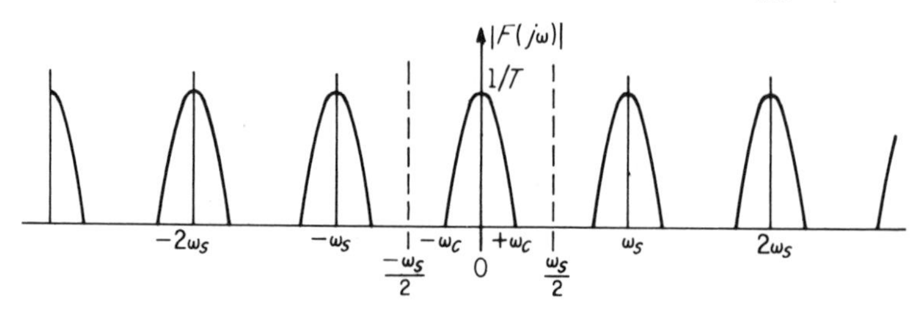

(b) Amplitude spectrum of sampler output.

FIG. 5-2. Amplitude spectra of input and output signals of an ideal sampler; $\omega_c < \frac{\omega_s}{2}$. (Reprinted by permission from Benjamin C. Kuo, *Analysis and Synthesis of Sampled-Data Control Systems*, Prentice-Hall, Inc. Englewood Cliffs, N.J. 1963, p. 35.)

†By bandlimited we mean that

$$|F(j\omega)| = 0 \quad \text{for} \quad \omega > \omega_c \quad \text{and} \quad \omega < -\omega_c$$

where ω_c is the cutoff frequency.

tinuous signal $f(t)$ is repeated with period ω_s along the frequency axis. This example illustrates that one effect of the sampler is the introduction of high frequency components into the frequency spectrum.

Now consider the reconstruction of $f(t)$ from $f^*(t)$, or equivalently, the recovery of $F(j\omega)$ from $F^*(j\omega)$. As the frequency spectrum of $F^*(j\omega)$ over the interval $-\omega_c \leqslant \omega \leqslant \omega_c$ is identical to that of $F(j\omega)$, this recovery can be effected if all frequencies outside this region are eliminated. Conceptually, this could be accomplished by the ideal filter whose amplitude spectrum is shown in Fig. 5-3.

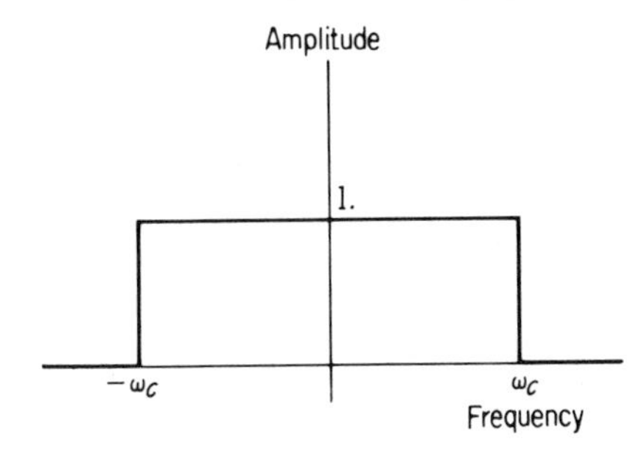

FIG. 5-3. Amplitude spectrum of an ideal filter.

Although practical filters can only approach the performance of the ideal filter, at least the frequencies of interest are not distorted in the sampled signal.

Now consider the case when the cutoff frequency is greater than half the sampling frequency, i.e., $\omega_c > \omega_s/2$. The amplitude spectrum of the continuous signal is shown in Fig. 5-4a. Again, the effect of the sampler is to cause this spectrum to be repeated along the frequency axis. However, in this case the part of the spectrum over the interval $\omega_s/2 < \omega < \omega_c$ is superimposed on the spectrum in the adjacent interval, giving the resulting amplitude spectrum Fig. 5-4b for $F^*(j\omega)$. We can see that parts of the spectrum in adjacent intervals are "folded over" on neighboring spectra, giving what is called the *foldover effect.*

Now consider reconstructing the spectrum of the continuous signal $|F(j\omega)|$ from the spectrum in Fig. 5-4b. If the ideal filter in Fig. 5-3 is used, the output spectrum (between $-\omega_s/2$ and $+\omega_s/2$ in Fig. 5-4b) would not match the amplitude spectrum of the continuous signal in Fig. 5-4a. Thus it is impossible in this case to recover the continuous signal.

These considerations are summarized by Shannon's sampling theorem: To be able to completely recover the continuous signal from its sampled counterpart, the sampling frequency ω_s must be at

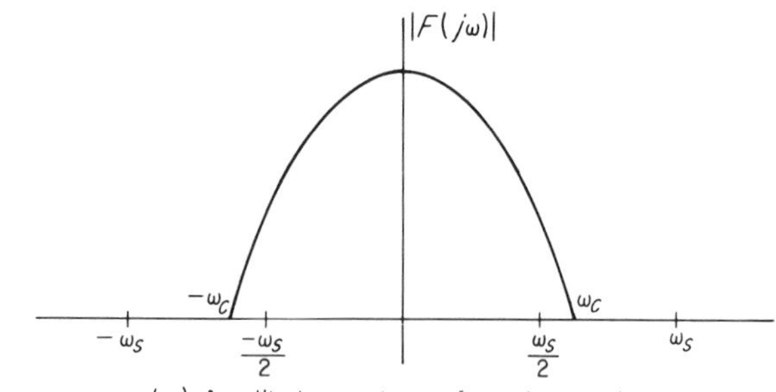

(a) Amplitude spectrum of continuous signal

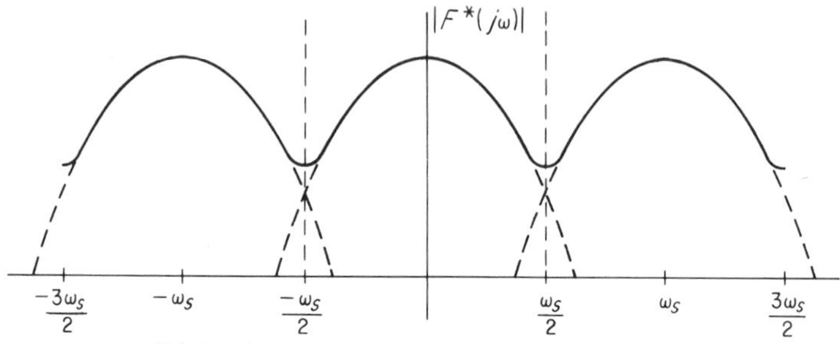

(b) Amplitude spectrum of sampled signal (Note: phase information is also necessary in order to add magnitudes)

FIG. 5-4. Amplitude spectra of input and output signals of an ideal sampler: $\omega_c > (\omega_s/2)$.

least twice the highest frequency (ω_c) in the signal. It should be noted that this is stronger than the statement, "the sampling frequency ω_s must be at least twice the maximum frequency to be recovered," a common misinterpretation. Unfortunately, most process signals are not bandlimited—that is, they have energies at all frequencies. In such cases, the continuous signal cannot be recovered from its sampled counterpart.

This also explains the "blow-up" phenomena in experimentally determined Bode plots for model development (2).

5-2 DATA HOLDS

The function of the hold element following the output multiplexer in Fig. 4-1 is to reconstruct a continuous signal, say $m(t)$, from a sequence of numbers, say m_n. In practice this device is often an integral part of the valve or other final control element. At least three points must be considered in the selection of the hold element:

1. Its ability to reconstruct the continuous signal. That is, suppose a continuous signal is sampled to produce a sequence of

numbers, which is in turn the input to the hold element. The question now becomes how well the reconstructed signal matches the original continuous signal.

2. Its effect on the performance of the control loop. This must be considered in two parts:
   (a) Its frequency response characteristics.
   (b) Its relationship to the control algorithm.
3. Practical aspects related to its construction.

Let us first consider the data reconstruction problem.

Specifically, consider the reconstruction of $m(t)$ over the interval $nT \leqslant t < (n+1)T$, or between the nth and $(n+1)^{\text{th}}$ samples. To be of practical utility in a control loop, this must be accomplished using only those values of the sequence m_n up to and including the nth entry. Otherwise, the data hold is not physically realizable.

One approach begins with the Taylor series expansion of $m(t)$ about the sampling instant nT:

$$m(t) = m(nT) + \frac{dm(nT)}{dt}(t - nT) + \frac{d^2 m(nT)}{dt^2}\frac{(t - nT)^2}{2} + \ldots, \quad nT \leqslant t < (n+1)T$$

Since $m(t)$ is a discontinuous function at the sampling instants, the derivative terms must in turn be expressed in terms of values of the sequence m_n up to and including the value at the sampling instant nT. This is readily accomplished using backward differences:

$$\frac{dm(nT)}{dt} = \frac{m(nT) - m[(n-1)T]}{T}$$

$$\frac{d^2 m(nT)}{dt^2} = \frac{\dfrac{dm(nT)}{dt} - \dfrac{dm[(n-1)T]}{dt}}{T} = \frac{m(nT) - 2m[(n-1)T] + m[(n-2)T]}{T^2}$$

and so forth.

Various holds can be proposed for approximating the Taylor series through a given number of terms. In fact, the order of the hold will be the order of the highest derivative in the terms approximated by the hold. Although the data reconstruction ability of the hold improves as more terms are approximated, there are two offsetting factors:

1. The use of values of m_n prior to the value at nT effectively requires a delay to be inserted into the control loop. This

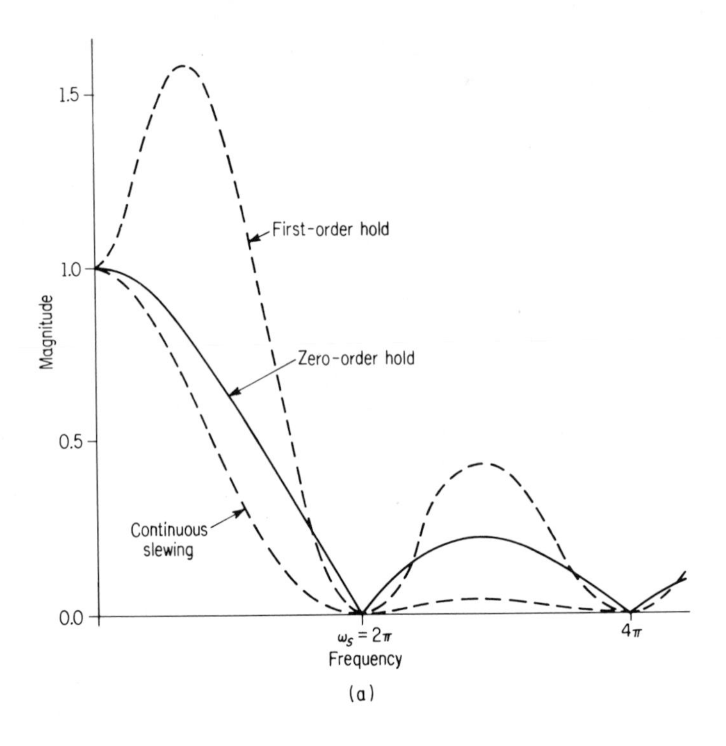

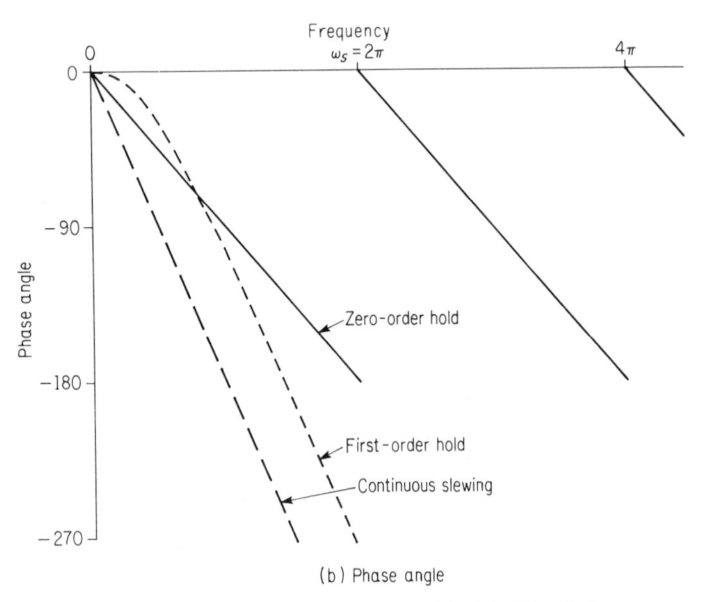

FIG. 5-5. Frequency response of holds ($T = 1.0$).

adversely affects the performance of the loop, imparting additional phase shift in the frequency response.

2. The physical complexity and cost of the hold increases with the order.

We shall now consider various holds.

Zero-Order Hold

Retaining only the first term in the Taylor series gives the following equation for the zero-order hold:

$$m(t) = m(nT) \qquad nT \leqslant t < (n+1)T$$

This hold was considered in the last chapter and its transfer function was derived to be

$$H_0(s) = \frac{1 - e^{-sT}}{s}$$

Its frequency response is consequently given by

$$H_0(j\omega) = \frac{1 - e^{-j\omega T}}{j\omega}$$

and is plotted in Fig. 5-5. Considering the ideal fitter to be the desired performance objective, the zero-order hold falls short in two regions:

1. The zero-order hold begins to significantly attenuate at frequencies considerably below ω_s.
2. The zero-order hold allows high frequencies to pass through although they are attenuated.

First-Order Hold

This hold is obtained by truncating the Taylor series after two terms:

$$m(t) = m(nT) + \frac{m(nT) - m[(n-1)T]}{T}(t - nT)$$

$$nT \leqslant t < (n+1)T$$

The output of this hold for a typical input is shown in Fig. 5-6b. Its transfer function $H_1(s)$ can be derived by taking the Laplace transform of its impulse response in Fig. 5-6c.

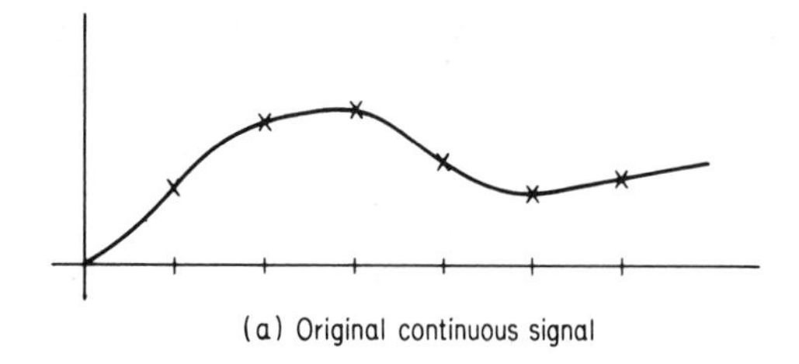

(a) Original continuous signal

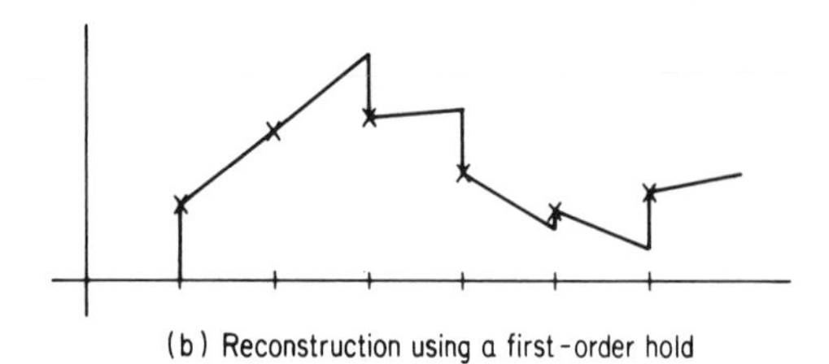

(b) Reconstruction using a first-order hold

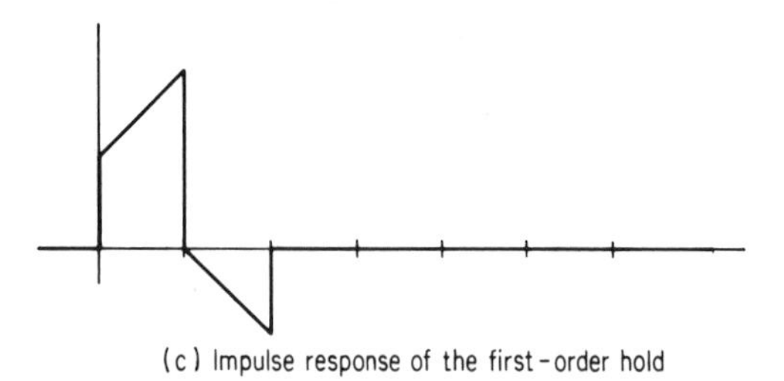

(c) Impulse response of the first-order hold

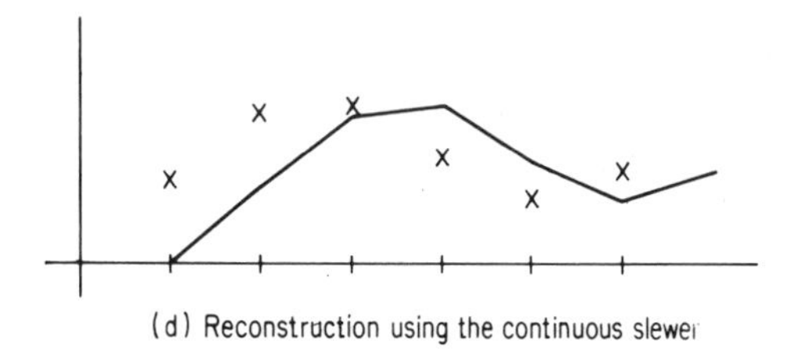

(d) Reconstruction using the continuous slewer

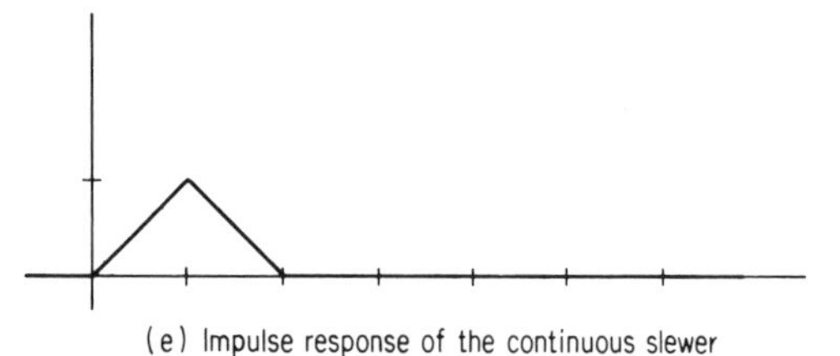

(e) Impulse response of the continuous slewer

FIG. 5-6. Performance of various holds.

$$H_1(s) = \mathcal{L}\left\{u(t) + \frac{t}{T}u(t) - 2u(t-T) - 2\frac{t-T}{T}u(t-T) + u(t-2T) + \frac{t-2T}{T}u(t-2T)\right\}$$

$$= \frac{1}{s} + \frac{1}{Ts^2} - \frac{2}{s}e^{-Ts} - \frac{2}{Ts^2}e^{-Ts} + \frac{e^{-2Ts}}{s} + \frac{e^{-2Ts}}{Ts^2}$$

$$= \frac{1+Ts}{T}\left(\frac{1-e^{-Ts}}{s}\right)^2$$

Its frequency response is given by

$$H_1(j\omega) = \frac{1+j\omega T}{T}\left(\frac{1-e^{-j\omega T}}{j\omega}\right)^2$$

The amplitude and angle plots are also shown in Fig. 5-5. This hold amplifies some frequencies in the low regions in addition to passing some high frequencies.

A second important feature of the first-order hold can be appreciated by comparing its phase angle plot with that of the zero-order hold. Note that at frequencies approaching the sampling frequency, the first-order hold imparts a greater phase shift than the zero-order hold. This is reflected by loops with a first-order hold having a lower ultimate gain than the same loops having a zero-order hold.

The fractional order hold is a modification of the first-order hold and is considered in the exercises.

Continuous Slewing

To physically implement either of the above two holds requires the actuator to change from one position to another quite rapidly. To avoid this demand on the actuator, a technique known as continuous slewing has been suggested (3). This is similar to the zero-order hold except that the change in position of the actuator is over the entire sampling interval rather than instantaneously. The output of this hold to a typical input is shown in Fig. 5-6d, and its impulse response in Fig. 5-6e. Its transfer function $H_c(s)$ is

$$H_c(s) = \mathcal{L}\left\{\frac{t}{T}u(t) - \frac{2(t-T)}{T}u(t-T) + \frac{t-2T}{T}u(t-2T)\right\}$$

$$= \frac{1}{Ts^2}(1 - 2e^{-Ts} + e^{-2Ts}) = \frac{1}{T}\left(\frac{1-e^{-sT}}{s}\right)^2$$

Its frequency response is given by

$$H_c(j\omega) = \frac{1}{T}\left(\frac{1 - e^{-j\omega T}}{j\omega}\right)^2$$

and is shown graphically in Fig. 5-5. Note that its phase lag is largest of all, but attenuates most at high frequencies.

5-3 DEAD TIME APPROXIMATION OF SAMPLE AND HOLD (4)

Consider the block diagram in Fig. 5-7 of a typical control loop found in the process industry. The loop consists of a feedback-con-

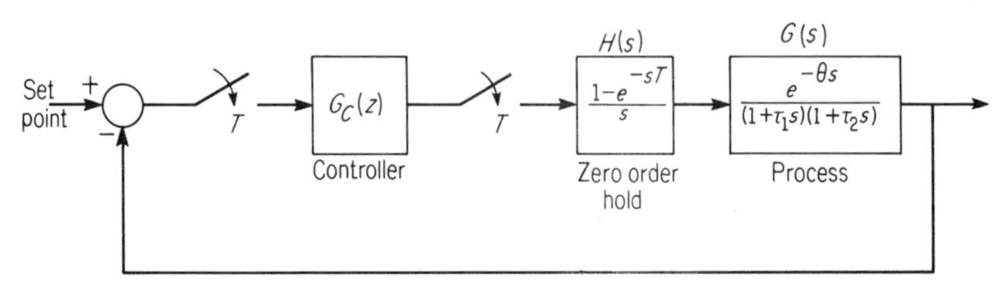

FIG. 5-7. Block diagram of a typical digital control loop with a sample and zero-order hold. (Reprinted by permission from Moore, Smith, and Murrill, "Simplifying Digital Control Dynamics for Controller Tuning and Hardware Lag Effects," *Instrument Practice*, January 1969.)

trol algorithm (e.g., proportional plus integral) and a process described by a second-order-lag plus dead time. The problem is to find a continuous transfer function $G_a(s)$ which approximates the dynamic behavior of the sample-and-hold device located between the controller and the plant in the discrete system (Fig. 5-7). This would effectively reduce the sampled-data loop to an equivalent continuous loop (Fig. 5-8).

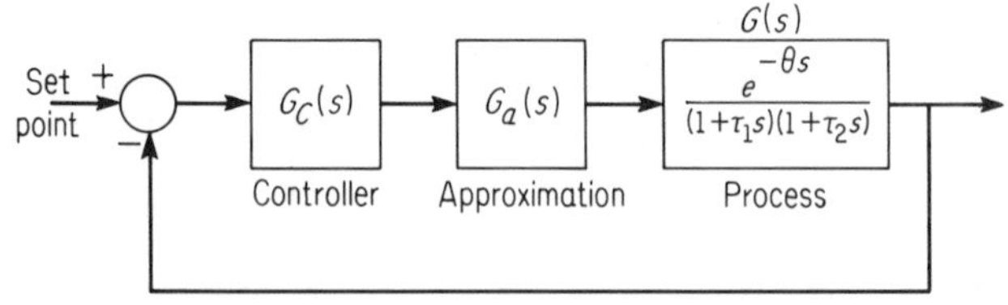

FIG. 5-8. Block diagram of an equivalent continuous control loop in which $G_a(s)$ approximates the behavior of the digital interface. (Reprinted by permission from Moore, Smith, and Murrill, "Simplifying Digital Control Dynamics for Controller Tuning and Hardware Lag Effects," *Instrument Practice*, January 1969.)

Intuitively an inspection of the response of the sample-and-hold device provides a clue to one possible approximation. Figure 5-9 illustrates the response of the sampler and zero-order hold to an arbitrary input. In most cases the hold is followed by a process with a large time constant which effectively smooths or averages the output. Note that the "average" output of the hold (shown by the dashed line in Fig. 5-9) is simply the original signal delayed by one-

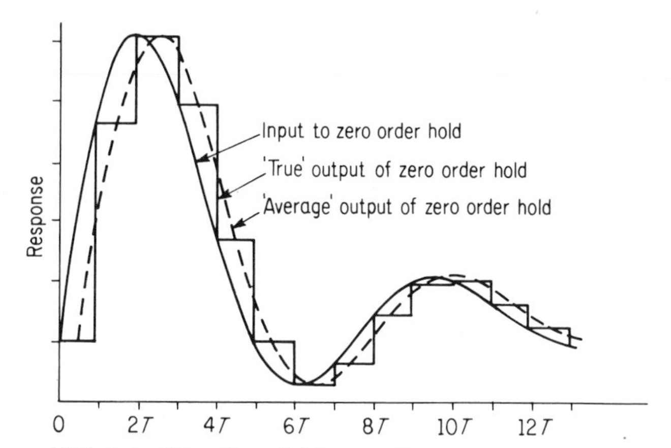

FIG. 5-9. "True" and "Average" response of a zero-order hold to an arbitrary input. (Reprinted by permission from Moore, Smith, and Murrill, "Simplifying Digital Control Dynamics for Controller Tuning and Hardware Lag Effects," *Instrument Practice*, January 1969.)

half the sampling time. When the "average" output of the hold is representative of the actual output, it appears that a pure time delay (transportation lag) of one-half the sampling time would adequately describe the behavior of the interface (i.e., $G_a(s) = e^{-(T/2)s}$).

Considering the behavior of the sample-and-hold device in the frequency domain provides some insight into the strengths and weaknesses of the dead time approximation. The open-loop transfer function for a sampler and zero-order hold followed by an arbitrary process can be described as follows.

$$HG^*(j\omega) = \frac{1}{T} \sum_{n=-\infty}^{\infty} H(j\omega + jn\omega_s)\, G(j\omega + jn\omega_s)$$

The $H(j\omega + jn\omega_s)$ term can be evaluated by considering the Laplace transfer function for the zero-order hold circuit:

$$H(s) = \frac{1 - e^{-sT}}{s}$$

Substituting $s = j\omega$, the transfer function can be written

$$H(j\omega) = \frac{2e^{-j\omega T/2}\ (e^{+j\omega T/2} - e^{-j\omega T/2})}{2j\omega}$$

$$= \frac{T \sin(\omega T/2)\, e^{-j\omega T/2}}{\omega T/2}$$

Therefore

$$H(j\omega + jn\omega_s) = \frac{T \sin\left[\left(\frac{\omega + n\omega_s}{2}\right) T\right]}{\left(\frac{\omega + n\omega_s}{2}\right) T}\ e^{-j(\omega + n\omega_s)T/2}$$

Since $T = \dfrac{2\pi}{\omega_s}$,

$$HG^*(j\omega) = \frac{1}{T} \Sigma \frac{T \sin\left(\pi \frac{\omega}{\omega_s} + n\pi\right)}{\left(\pi \frac{\omega}{\omega} + n\pi\right)}\ e^{-j(\omega + n\omega_s)T/2}\, G(j\omega + jn\omega_s)$$

When the process $G(s)$ is insensitive to high-frequency terms, this expression reduces to

$$HG^*(j\omega) \cong \frac{\sin \pi \frac{\omega}{\omega_s}}{\pi \frac{\omega}{\omega_s}}\ e^{-j\omega T/2}\, G(j\omega),\ |\omega| < \omega_s$$

That is, all terms except the one corresponding to $n = 0$ can be neglected. Furthermore, when ω/ω_s is small,

$$\frac{\sin \pi \frac{\omega}{\omega_s}}{\pi \frac{\omega}{\omega_s}} \cong 1$$

and

$$HG^*(j\omega) \cong e^{-\omega T/2}\, G(j\omega)$$

or

$$HG^*(s) \cong e^{-sT/2}\,G(s)$$

That is, the sample and hold behaves much like a pure dead time equal to half the sampling time.

On the surface the assumptions necessary in this development do not appear to be severe for the typical process operation. In general, most plants found in the process industries are low-pass filters by nature, and therefore they are not sensitive to the high-frequency terms generated by the sampler ω_s. Also, under most operating conditions the fundamental frequency of the system response ω will be lower than the sampling frequency (ω_s); therefore, ω/ω_s will usually be small.

To substantiate the validity of the above assumptions and evaluate the actual performance of the approximation, a typical plant was selected and studied both in the frequency and in the time domain. Figures 5-10 and 5-11 are the results of a study conducted with the second-order plant,

$$G(s) = \frac{1.0}{(\tau s + 1)^2}$$

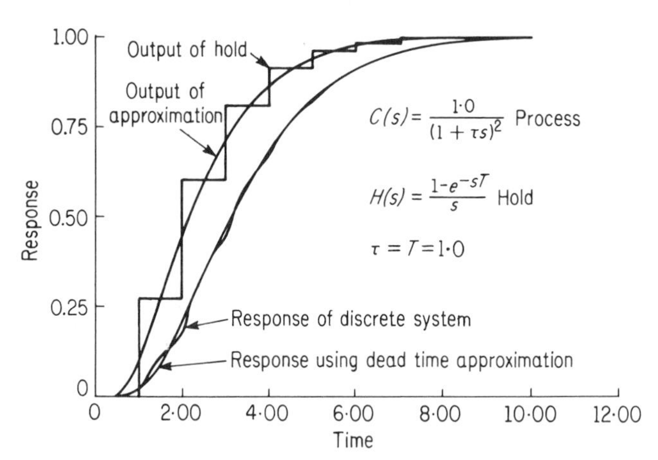

FIG. 5-10. Comparison of open-loop response of a discrete system with a sampler and zero-order hold to the continuous system using the dead-time approximation. (Reprinted by permission from Moore, Smith, and Murrill, "Simplifying Digital Control Dynamics for Controller Tuning and Hardware Lag Effects," *Instrument Practice*, January 1969.)

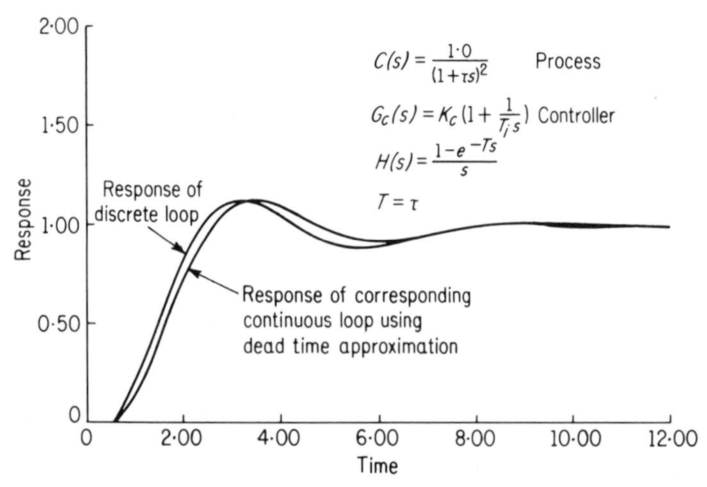

FIG. 5-11. Comparison of closed-loop response of a discrete system with a sample and zero-order hold to the continuous system using a dead-time approximation. (Reprinted by permission from Moore, Smith, and Murrill, "Simplifying Digital Control Dynamics for Controller Tuning and Hardware Lag Effects," *Instrument Practice*, January 1969.)

where τ = time constant of the plant. Illustrated are the open-loop response, closed-loop response (using a PI controller), and frequency response for the case $\tau = 1.0$.

In Figs. 5-10 and 5-11 the open-loop and closed-loop response are illustrated at a rather large sampling time ($T = \tau = 1.0$). Only this extreme value is shown because at lower, more reasonable sampling times the response of the actual discrete system and the response of the continuous approximation are essentially indistinguishable. Part of the reason the time response comparisons look so good can be understood by looking at the frequency response plots shown in Fig. 5-12. The frequency-response curves for this system essentially show that the primary factor determining the effectiveness of the approximation is the "fold-over" frequency (predicted by Shannon's sampling theorem) which says that the sampling rate must be at least twice the highest frequency component of the response. If sampling is below this frequency, the frequency response "folds over" and becomes distorted as is vividly demonstrated in Fig. 5-12. Below this critical frequency the difference between the discrete system and the continuous approximation is very low in both the magnitude plot and in the phase plot. Above the critical frequency the difference is quite large as would be expected.

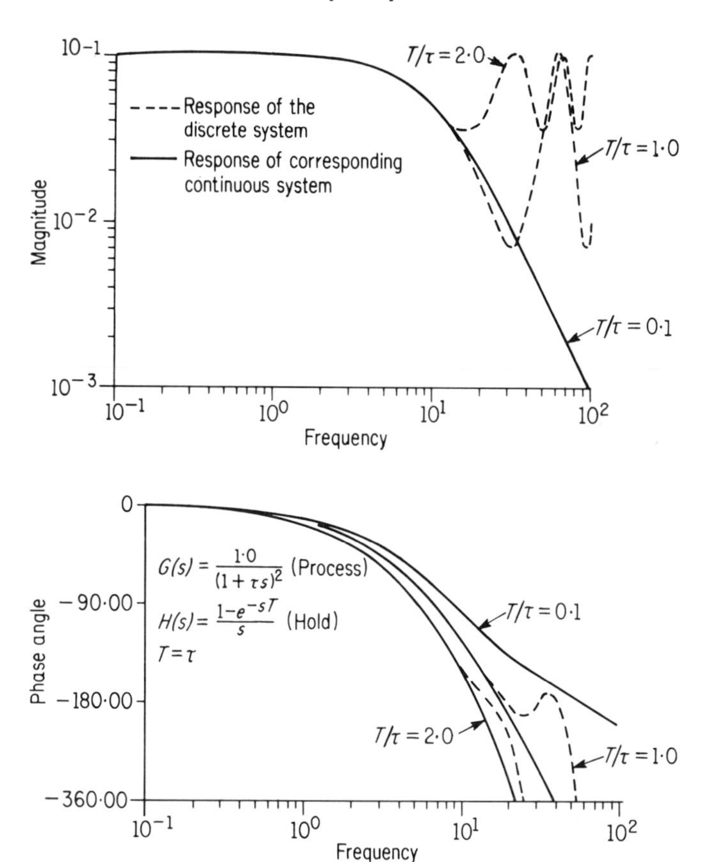

FIG. 5-12. Comparison of frequency response of a discrete system with a sample and zero-order hold to the continuous system using a dead-time approximation. (Reprinted by permission from Moore, Smith, and Murrill, "Simplifying Digital Control Dynamics for Controller Tuning and Hardware Lag Effects," *Instrument Practice*, January 1969.)

5-4 *s*-PLANE CHARACTERISTICS OF SAMPLED FUNCTIONS

The development in Sec. 5-1 of the periodic nature of $F^*(j\omega)$ can also be applied to $F^*(s)$. Recall that the Laplace transform of $f^*(t)$ is

$$F^*(s) = \sum_{n=0}^{\infty} f(nT)\, e^{-nTs}$$

Suppose we now consider $F^*(s + j\omega_s)$, which is

$$F^*(s + j\omega_s) = \sum_{n=0}^{\infty} f(nT)\, e^{-nT(s+j\omega_s)}$$

$$= \sum_{n=0}^{\infty} f(nT)\, e^{-nTs}\, e^{-jnT\omega_s}$$

But since $e^{-jnT\omega_s}$ is unity, this expression reduces to

$$F^*(s + j\omega_s) = \sum_{n=0}^{\infty} f(nT)\, e^{-nTs} = F^*(s)$$

This relationship can be generalized to

$$F^*(s + jm\omega_s) = F^*(s), \qquad m = 0, \pm 1, \pm 2, \ldots \tag{5-4}$$

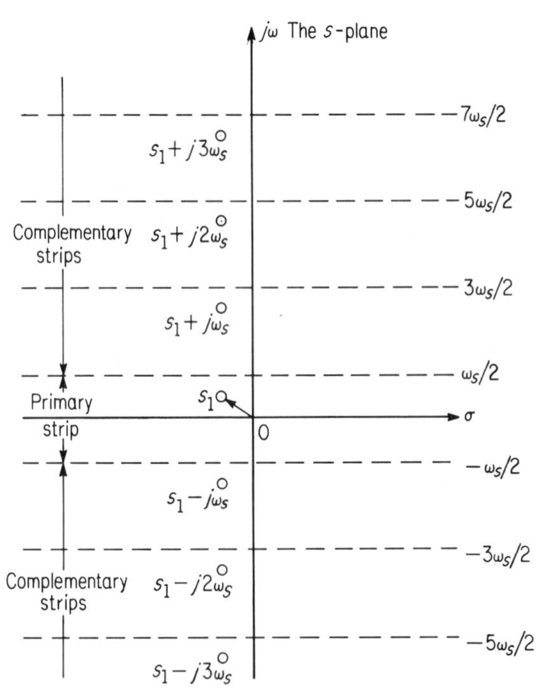

FIG. 5-13. Periodic strips in the s-plane. (Reprinted by permission from Benjamin C. Kuo, *Analysis and Synthesis of Sampled-Data Control Systems*, Prentice-Hall, Inc., Englewood Cliffs, N.J., 1963, p. 36.)

Thus $F^*(s)$ is a periodic function in the complex plane with period $j\omega_s$. This means that the s-plane can be divided into strips of width ω_s along the imaginary axis as shown in Fig. 5-13. The strip from $-j\omega_s/2$ to $+j\omega_s/2$ is called the *primary strip*, all others being repetitions of the strip and called *complementary strips.*

It follows from Eq. 5-4 that if a pole or zero occurs anywhere in the primary strip, one must be located at the corresponding point in each of the complementary strips. That is, if $F^*(s)$ has a zero at s_1, there is also a zero at $s = s_1^*$ and at corresponding points in the other strips. This is also illustrated in Fig. 5-13.

5-5 STABILITY IN THE z-DOMAIN

In the preceding chapter, we concluded that for a sampled data system to be stable, all poles must lie within the unit circle except for possibly one at $z = 1$. In this section we shall arrive at this conclusion via another route.

Recall that in order for a system to be stable, all poles must be

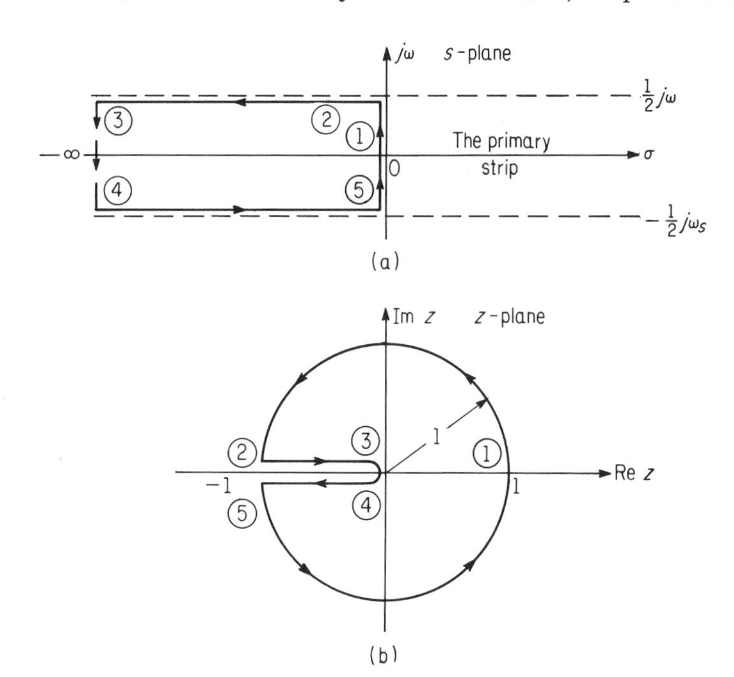

FIG. 5-14. Mapping of the primary strip in the left-half of the s-plane into the z-plane. (Reprinted by permission from Benjamin C. Kuo, *Analysis and Synthesis of Sampled-Data Control Systems*, Prentice-Hall, Inc., Englewood Cliffs, N.J., 1963, p. 59.)

in the left-hand s-plane except for possibly one at the origin. As the Laplace transform of a sampled function is periodic with period $j\omega_s$, only the primary strip, not the entire s-plane, needs to be examined. For the primary strip in Fig. 5-14, the system would be stable provided the closed path as shown encloses all the poles lying in the primary strip, i.e., no poles in the right half of the primary strip.

The next step is to relate the stability criterion in the s-plane to a stability criterion in the z-plane by mapping the path in Fig. 5-14a into the z-plane. This is readily accomplished by the relationship $z = e^{Ts}$. For example, the path from (1) to (2) is given by

$$s = j\omega \qquad 0 \leqslant \omega \leqslant j\omega_s/2$$

In the z-domain, this is

$$z = e^{j\omega T} = 1.0 \angle \omega T \qquad 0 \leqslant \omega \leqslant j\omega_s/2$$

or

$$z = 1.0 \angle \alpha \qquad 0 \leqslant \alpha \leqslant 180^\circ$$

This is the upper half of the circle in Fig. 5-14b. This type of reasoning can be continued to obtain the entire path in Fig. 5-14b. In order for a system to be stable, all poles must lie inside this path except for possibly one at $z = 1.0$.

5-6 NOISE REJECTION (3)

Since the signals emanating from the typical industrial process have noise superimposed upon the useful information from which meaningful control action can be determined, it is necessary to incorporate noise-rejection schemes into our control loops, as illustrated in Fig. 5-15. Note that a two-stage analog/digital filtering scheme is shown. While it is conceptually possible to use only an analog filter, this approach suffers some disadvantages that we shall enumerate shortly.

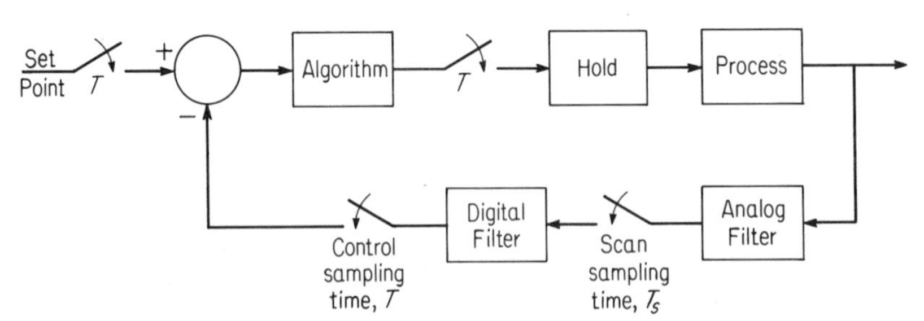

FIG. 5-15. Control loop including noise-rejection schemes.

Pure digital filtering is not recommended due to the aliasing or fold-over phenomenon. Figure 5-16 illustrates that when a high-frequency sinusoidal is sampled at 7/8 samples per cycle, a low-frequency signal is produced whose amplitude equals that of the original signal. As this occurs at the scan sampler, digital filtering cannot avoid this. Thus an analog filter is generally needed to remove high-frequency noise prior to the scan sampler.

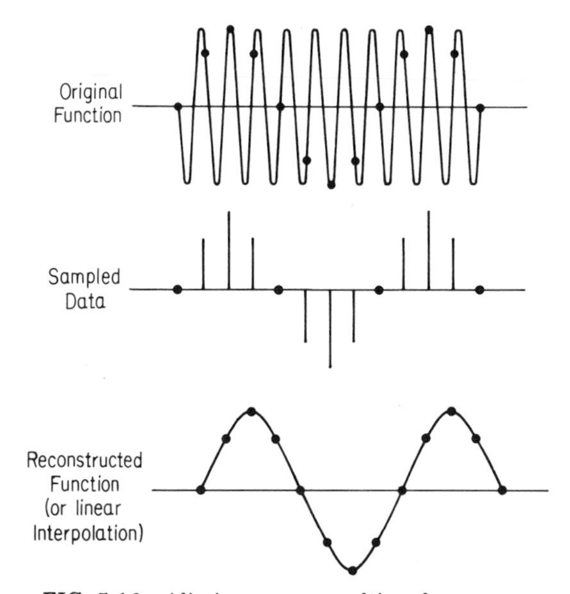

FIG. 5-16. Aliasing error resulting from sampling at rates of 7/8 samples per cycle or original function. (Reprinted by permission from Ref. 3.)

The sources of noise are numerous and obvious, ranging from high-frequency a-c components to low-frequency process fluctuations. As shown in Fig. 5-17, the input signal is composed of (1) controllable disturbances, (2) uncontrollable disturbances, (3) measurement noise, and (4) stray electrical pickup. These latter three are all classified as noise. In cases in which the noise in the input signal is not appreciable, the only filtering is with an analog RC network in the analog front end of the computer. As illustrated by case I in Fig. 5-17, the time constants 1/RC of these filters are such as to only eliminate the high-frequency disturbances. The frequency aspects of these filters are those of the first-order lag as shown in Fig. 5-18.

When appreciable noise is present at lower frequencies, the rejection may be accomplished either by (1) all analog filtering as in case II in Fig. 5-17, or (2) by combined analog and digital filter rejection as

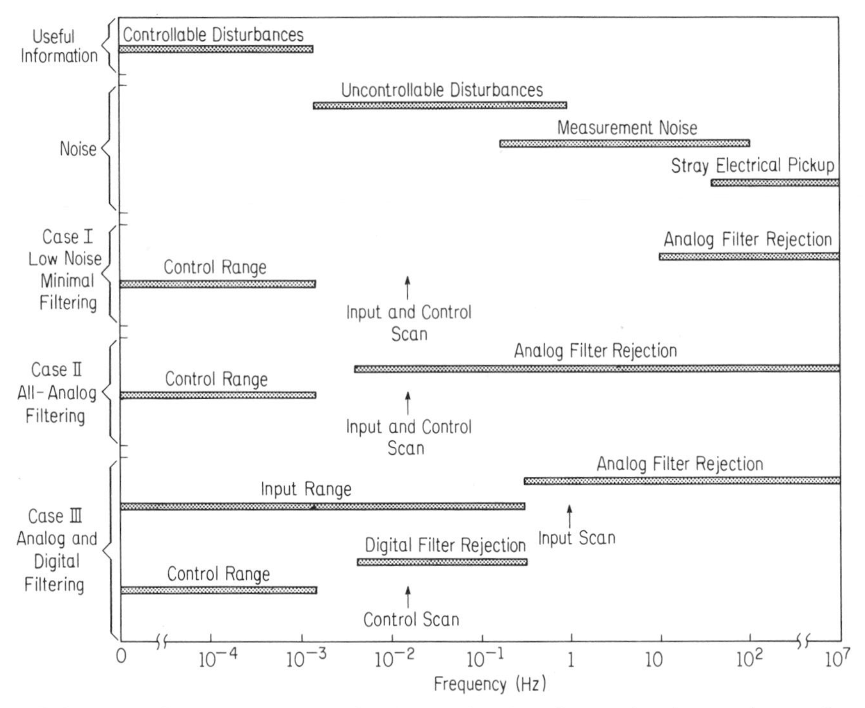

FIG. 5-17. Frequency ranges for input signal and associated scanning and filtering functions for a typical process. (Reprinted by permission from Ref. 3.)

in Case III. The pertinent factors to consider in this selection are

1. Analog filters composed only of passive elements (resistors, capacitors, etc.) are practical only for time constants up to a few seconds. For larger time constants, active elements (operational amplifiers) are required.
2. The time constant of a digital filter is readily changed from the computer console, whereas the same change for an analog filter could require electrical modifications.
3. To use digital filtering, the input scan rate must usually be increased (compare Cases II and III in Fig. 5-17), which increases the load on the computer.
4. Digital filtering requires some computational time and storage whereas analog filtering requires none.

These considerations seem to make Case II in Fig. 5-17 the most attractive.

In all cases in Fig. 5-17 the filter rejection does not extend all the way to the control range. This arises from the fact that the first-

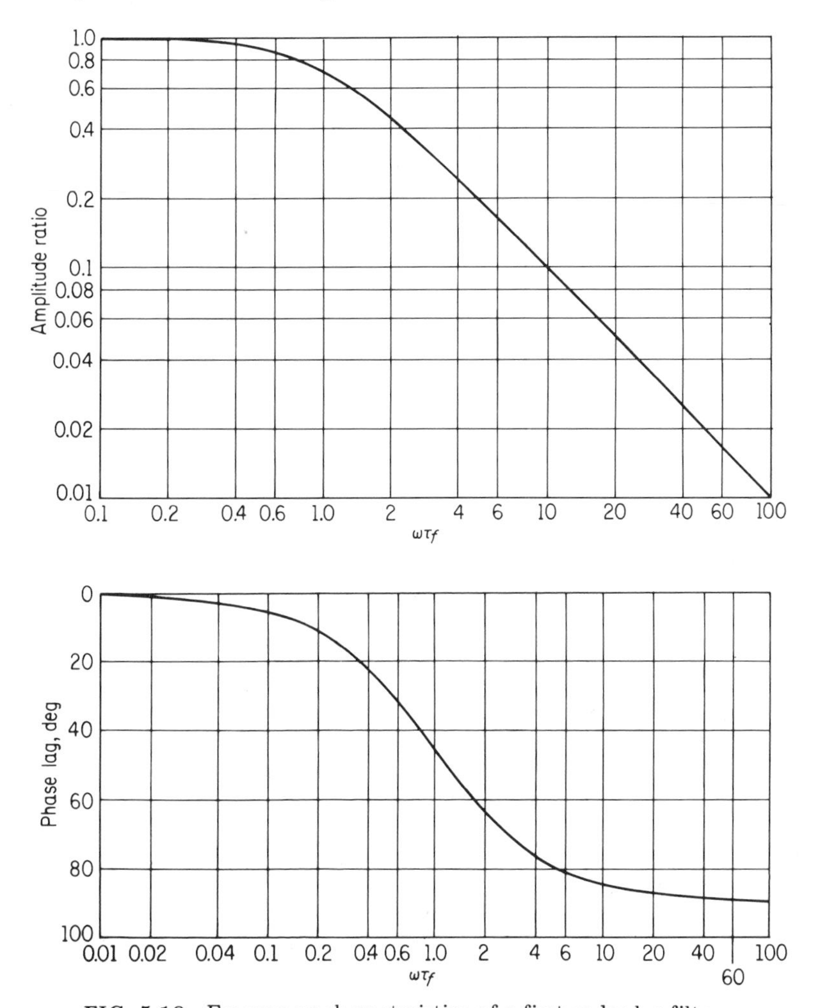

FIG. 5-18. Frequency characteristics of a first-order lag filter.

order lag does not have a sharp cutoff point, and the gap is provided to avoid alteration in the control range. Although higher-order filters with sharper cutoffs can be designed, they have not enjoyed widespread use in process computer control systems.

Suppose a reading, say x_i, is taken every T_s units of time. From these readings, two algorithms are commonly used to calculate the filtered or smoothed values y_i.

The first algorithm is simply the arithmetic average

$$y_n = \frac{1}{N} \sum_{i=0}^{N-1} x_{n-i}$$

Since N is the number of readings for x_i required to calculate one value of y_n, the control action can be taken only at intervals equal to $NT_s = T$ units of time.

The second algorithm is the numerical equivalent of the first-order lag. For continuous signals, the differential equation describing the first-order lag is

$$\tau \frac{dy(t)}{dt} + y(t) = x(t)$$

Expressing this equation by finite differences,

$$\tau \left(\frac{y_n - y_{n-1}}{T_s} \right) + y_{n-1} = x_n$$

or

$$y_n = \alpha x_n + (1 - \alpha) y_{n-1} \tag{5-5}$$

where $\alpha = T/\tau_f$. This relationship between α and τ_f is valid only when $\tau_f >> T$, but the general form of Eq. 5-5 is used for filtering even when this does not hold (see exercises for development of equation using z-transforms). Figure 5.19 gives the frequency characteristics of both filters.

The implementation of either of these algorithms is certainly feasible with regard to programming, computational requirements, and storage requirements.

Some of the beneficial effects of filtering can be appreciated from an examination of Fig. 5.20. The manipulated variable is $m(t)$, the controlled variable is $c(t)$, and the low-frequency part of $c(t)$ is denoted as $c'(t)$. In each case a PI controller was used and was retuned for each case. No filtering was used for Fig. 5.20a, and an analog filter with a time constant of 25 sec was used in Fig. 5.20b. Two items should be noted:

1. The amplitudes of the changes in $m(t)$ are lower for the filtered case. This reduces actuator wear.
2. The low-frequency variations in $c(t)$ are smaller for the filtered case. The explanation for this is that the high-

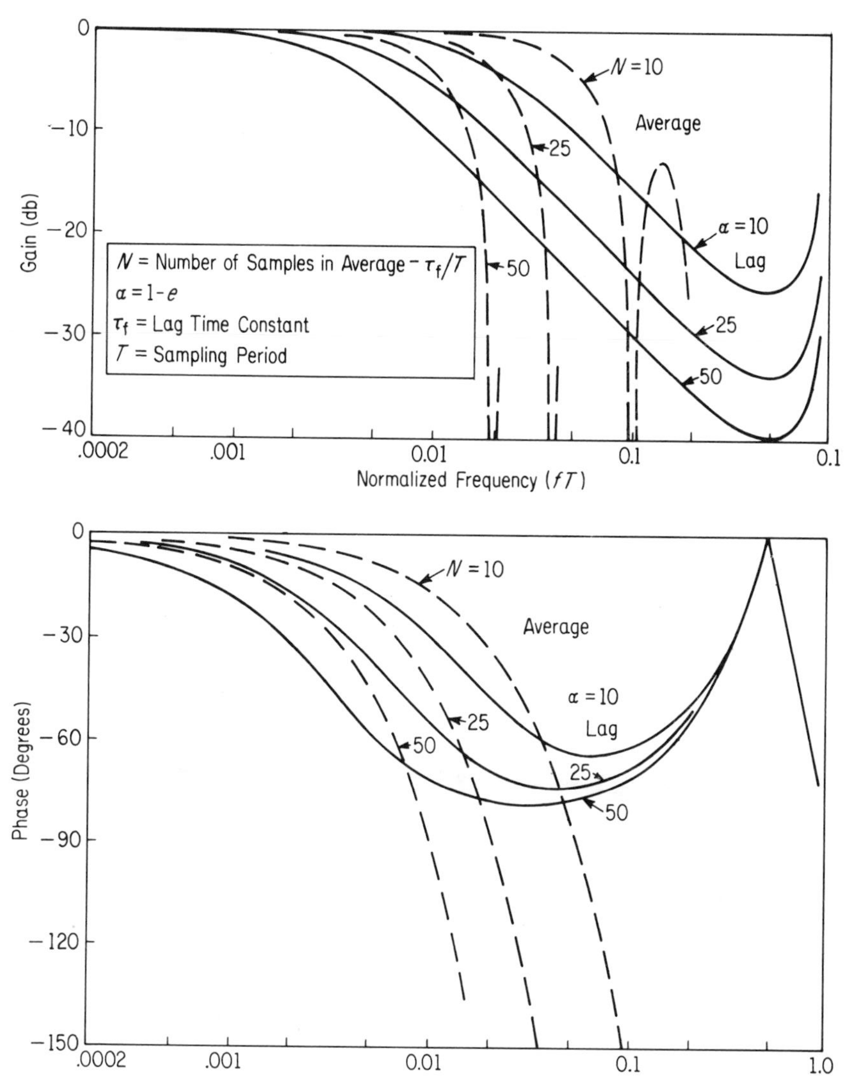

FIG. 5-19. Comparison of gain and phase characteristics for average and digital lag. (Reprinted by permission from Ref. 3.)

frequency components are removed by filtering, thus reducing the aliasing of high frequencies onto low frequencies.

A guide to the selection of the filter time constant can be obtained by considering the typical frequency spectrum in Fig. 5.21. For sampling at frequency ω_s with no prior filtering, the frequency components above $\omega_s/2$ are folded onto the low-frequency com-

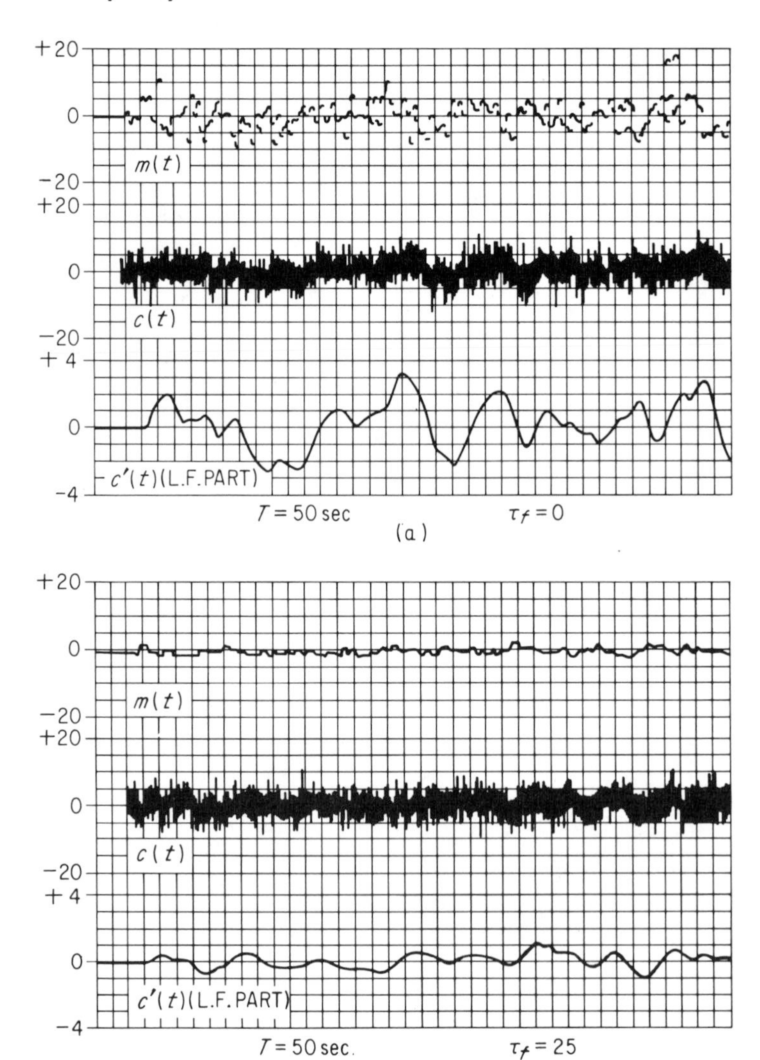

FIG. 5-20. Effect of noise on process variables for control with and without a filter. Each time division is 250 sec. (Reprinted by permission from Ref. 3.)

ponents. This could be avoided if all components at frequencies above $\omega_s/2$ were removed by filtering. Although all these components cannot be completely removed, use of a first-order lag with a corner frequency at $\omega_s/2$ would attenuate at frequencies above ω_s. The corner frequency of the first-order lag is simply the reciprocal of

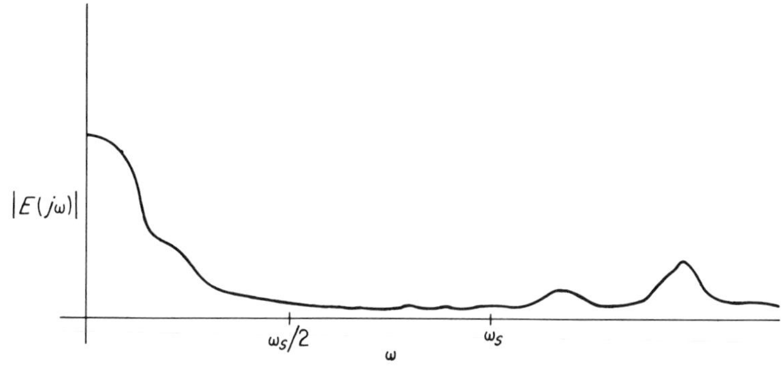

FIG. 5-21. Typical frequency spectrum of a process signal.

the time constant. Thus

$$\frac{1}{\tau_f} = \frac{\omega_s}{2} = \frac{\pi}{T}$$

and

$$\tau_f = T/\pi$$

A common rule-of-thumb is to select the filter time constant equal to $T/2$.

For the analog filter, its time constant should be about $T_s/2$, when T_s is the scan sampling time as was illustrated in Fig. 5-15. For the digital filter, the time constant should be $T/2$, where T is the control sampling time.

LITERATURE CITED

1. Kuo, B. C., *Analysis and Synthesis of Sampled-Data Control Systems*, Prentice-Hall, Englewood Cliffs, N.J., 1963.
2. Murrill, P. W., R. W. Pike, and C. L. Smith, "Pulse-Testing Methods," *Chemical Engineering*, Vol. 76, No. 4 (Feb. 24, 1969), pp. 105–108.
3. Goff, K. W., "Dynamics in Direct Digital Control," Part I, *ISA J.*, Vol. 13, No. 11 (November 1966), pp. 45-49; Part II, *ISA J.*, Vol. 13, No. 12 (December 1966), pp. 44-54.
4. Moore, C. F., C. L. Smith, and P. W. Murrill, "Simplifying Digital Control Dynamics for Controller Tuning and Hardware Lag Effects," *Instrument Practice*, Vol. 23, No. 1(January 1969), pp. 45–49.

chapter 6

Control Algorithms

The purpose of this chapter is to present techniques for developing control algorithms. We shall consider both z-transform design techniques and the use of conventional algorithms. To date, the latter type of algorithms have been used in the majority of digital control systems. The z-transform techniques have been utilized only when significant incentive exists for their use, e.g., processes with large dead time. While we shall surely see more use of the z-transform techniques, the conventional PI and PID algorithms are certainly not going to disappear.

In order to design a control algorithm using z-transforms, a process model is required. Although the PI and PID algorithms are frequently if not usually tuned by trial-and-error approaches, a model can be used to at least give initial estimates of the tuning parameters. For these reasons, the first section of this chapter is devoted to the development of simple process models using shortcut techniques. More rigorous methods are presented in Chapter 7. The models will be developed for a fourth-order system, and will be used to design control algorithms for this system in the remaining sections of this chapter and in the exercises.

6-1 SIMPLE PROCESS MODELS

The techniques presented in this section all have the following characteristics:

1. The parameters in the model are determined from the open-loop step response of the process.
2. Determination of the model parameters is via a graphical construction and/or simple calculations that can be readily accomplished with a slide rule; i.e., a computer is not required.

Because of the latter feature, the accuracy of models so determined is not generally the best possible. The techniques presented in the next chapter are more accurate, but also require more computations.

To begin our discussion, consider what general model form would be appropriate to most process systems. If a step change is made in the manipulated input to a system such as a temperature process, the process will seek a new level. Systems of this type are said to be self-regulating, and their dynamics can be described by one or more time constants. But if a step change is made in the manipulated input to systems such as many level control processes, the process output (i.e., tank level) will not seek a new level, but will increase or decrease indefinitely. Systems of this type are said to be nonself-regulating, and their dynamics must contain at least one integral term and possibly one or more time constants.

For nonself-regulating processes, a typical response is shown in Fig. 6-1. The parameters in the model

$$G_m(s) = \frac{Ke^{-\theta s}}{s}$$

can be readily determined graphically by first drawing a line tangent

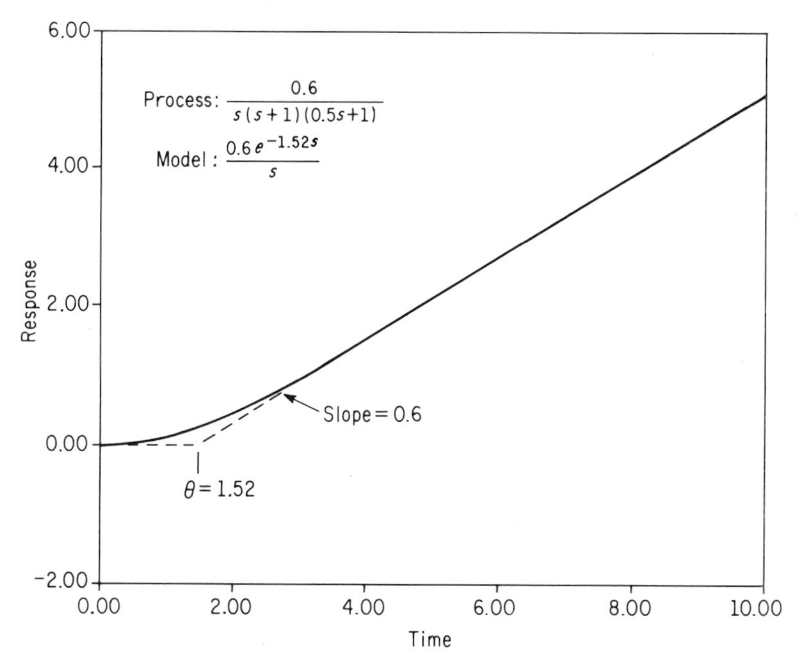

FIG. 6-1. Derivation of a simple model for a nonself-regulated process.

to the response when the rate of rise becomes nearly constant, as illustrated in Fig. 6-1. The slope of this line is the gain K and the intercept with the base line is the dead time θ. This model illustrates the use of the dead time to approximate high-order time constants, of which we shall see more in subsequent examples. Many processes, however, also contain significant true dead time.

The simplest model for a self-regulating process contains a single time constant, giving the first-order-lag-plus-dead-time model:

$$G_m(s) = \frac{Ke^{-\theta s}}{\tau s + 1} \tag{6-1}$$

where K is the process gain, θ the dead time, and τ the time constant. If two time constants are to be used, the model becomes the second-order-lag-plus-dead-time model:

$$G_m(s) = \frac{Ke^{-\theta s}}{(\tau_1 s + 1)(\tau_2 s + 1)} \tag{6-2}$$

where τ_1 and τ_2 are the time constants. Alternatively, this model can be expressed in terms of a damping ratio ζ and a natural frequency ω_n:

$$G_m(s) = \frac{Ke^{-\theta s}}{\dfrac{s^2}{\omega_n^2} + 2\zeta \dfrac{s}{\omega_n} + 1} \tag{6-3}$$

Although processes whose open-loop step response exhibits an overshoot are rare, we shall subsequently see that some higher-order processes are best described by a second-order model with a damping ratio ζ slightly less than one.

If the complexity of the process model is to be increased still further, the next logical step would be to add a zero.

$$G_m(s) = \frac{Ke^{-\theta s}(s/\alpha + 1)}{(\tau_1 s + 1)(\tau_2 s + 1)}$$

where α is the zero. This type of model is necessary in order to approximate processes whose step response is as shown in Fig. 6-2. Fortunately, these cases are not too common.

In order to understand the basis of the methods for obtaining a first-order-lag-plus-dead-time model, first consider the step response of an actual first-order-lag-plus-dead-time system as illustrated in Fig. 6-3a. Note that the response attains 63.2 percent of its final value in one time constant. Also, note that a line drawn tangent to the response at the point where the rate of change or derivative is at

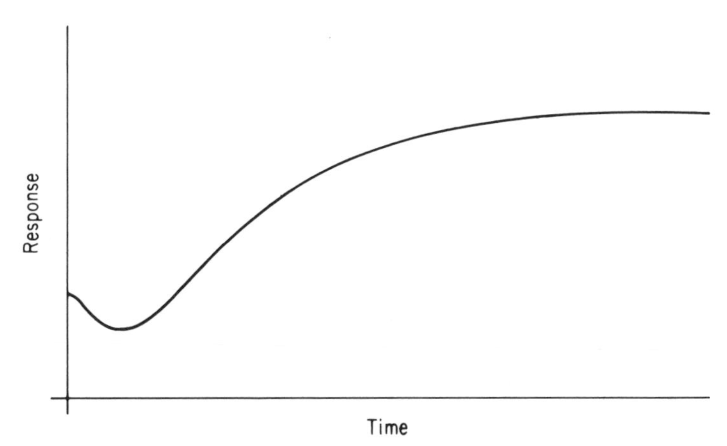

FIG. 6-2. Example of a step response that requires a second-order-lag-plus-first-order lead for a good approximation.

a maximum intersects the 100 percent response line at time equal to the time constant.

Using these characteristics of the first-order lag as a basis on which to develop a model, suppose we first draw a line tangent to the step response of the process (also called the *process reaction curve*). This is illustrated in Fig. 6-3b for the high-order process

$$G(s) = \frac{1}{(0.5s + 1)(s + 1)^2 \ (2s + 1)}$$

The model dead time θ is chosen as the point at which this tangent

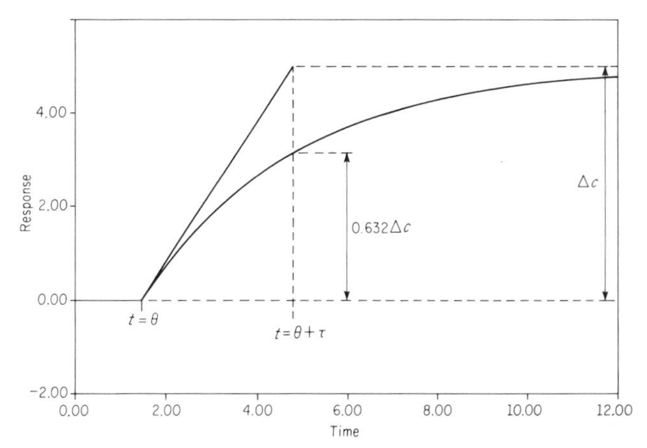

FIG. 6-3a. Response of an actual first-order-lag-plus-dead-time system.

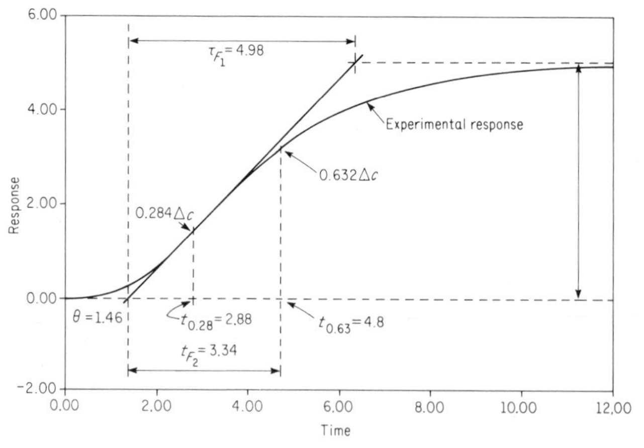

FIG. 6-3b. Construction used to determine the three first-order models (Δm = 5.0).

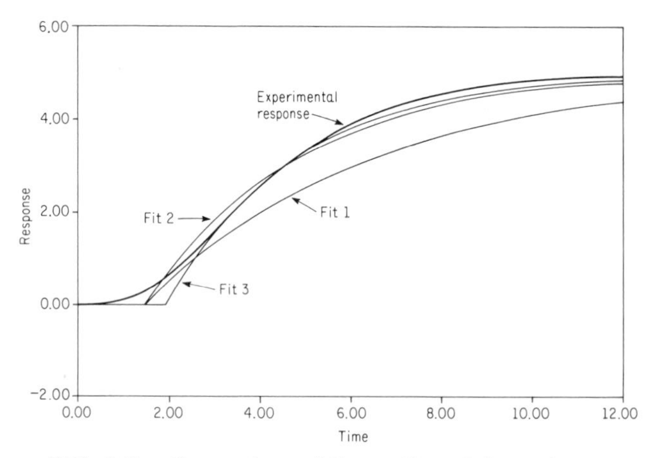

FIG. 6-3c. Comparison of the quality of the various first-order approximations.

line intersects the base line. The time constant can be determined in one of two ways:

1. The time at which the tangent line intersects the 100 percent response line equals $\theta + \tau$. This is essentially the method originally proposed by Ziegler and Nichols (1). We shall refer to the model parameters determined in this manner as θ_{F_1} and τ_{F_1} (Fit 1).
2. The time at which the process response attains 63.2 percent of the final value equals $\theta + \tau$. This modification was suggested by Miller (2), and the model parameters shall be referred to as θ_{F_2} and τ_{F_2} (Fit 2).

In either case, the model gain K is determined as the ratio Δc to the change in input Δm. The parameters determined by these two methods are shown in Table 6-1. An idea of the quality of these approximations can be judged from Fig. 6-3c. The Fit 2 parameters generally appear to give a better approximation than Fit 1.

TABLE 6-1
First-Order-Lag-Plus-Dead-Time Model Parameters

	Dead Time	Time Constant	Gain
Fit 1	1.46	4.98	1.0
Fit 2	1.46	3.34	1.0
Fit 3	1.92	2.88	1.0

Although drawing the tangent line is conceptually simple, it is difficult to accurately draw in practice. An alternate route to the approximation is to determine θ and τ from two points on the process reaction curve. The analytic solution for the step response of the first-order-lag-plus-dead-time model is

$$c(t) = (\Delta m)\,(1 - e^{-(t-\theta)/\tau}),\ t > \theta \qquad (6\text{-}4)$$

Suppose we calculate $c(t)$ at two times, say $\theta + \tau/3$ and $\theta + \tau$:

$$c(\theta + \tau/3) = 0.284(\Delta c) \qquad (6\text{-}5)$$

$$c(\theta + \tau) = 0.632(\Delta c) \qquad (6\text{-}6)$$

These two points were chosen quite arbitrarily, and others could certainly have been used.

The times that the response attains 28.4 percent and 63.2 percent of the final value can be read from the process response, as illustrated in Fig. 6-3b:

$$t_{0.28} = 2.88 = \theta + \tau/3 \qquad (6\text{-}7)$$

$$t_{0.63} = 4.80 = \theta + \tau \qquad (6\text{-}8)$$

These two equations can be solved simultaneously for θ and τ, which we shall designate as Fit 3 to distinguish from the previous models.

$$\theta_{F_3} = 1.92$$

$$\tau_3 = 2.88$$

The response of this model is also shown in Fig. 6-3c.

While a second-order model will give a better fit than a first-order model, determination of the additional model parameter is not especially easy. The first step is to draw the line tangent to the process reaction curve at its inflection point, noting the values of T_A and T_c

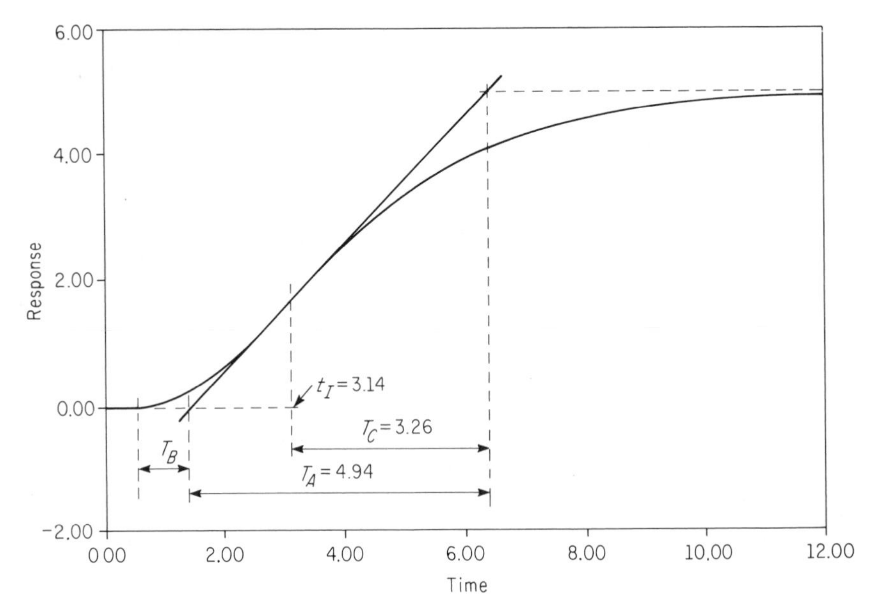

FIG. 6-4. Tangent line used to develop a second-order model.

as illustrated in Fig. 6-4. The ratio of the time constants is related to the ratio T_C/T_A by the equation

$$\frac{T_C}{T_A} = (1 + x)x^{x/(1-x)} \tag{6-9}$$

where x is τ_1/τ_2. The sum of the time constants is

$$\tau_1 + \tau_2 = T_C \tag{6-10}$$

with T_C and T_A known, these two equations can be solved implicitly for the time constants τ_1 and τ_2. The graph in Fig. 6-5 facilitates this. For each value of T_C/T_A, two solutions (one for each time constant) can be determined from curve A. This graph is taken from an article by Sten (3), although the original development can be attributed to Oldenbourg and Satorius (4).

To determine the dead time, Sten used Curve B in Fig. 6-5. From the value of T_A and T_C the value of T_B can be determined from curve B in Fig. 6-5, or from the following approximation:

$$\frac{T_B}{T_A} = -0.4729\,\frac{T_C}{T_A} + 0.4512 \tag{6-11}$$

The significance of T_B is that a second-order response curve passing through the inflection point with the proper slope would originate at

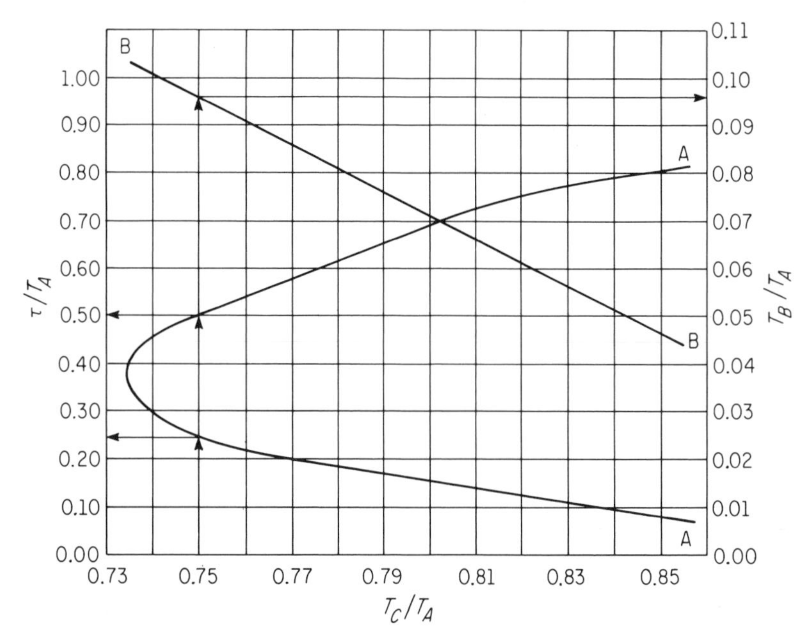

FIG. 6-5. Graphical solution of Eqs. 6-9 and 6-10. (Reproduced by permission from Ref. 3.)

the time t_I at the inflection point plus T_C minus the sum of T_A and T_B. Thus the dead time θ is given by

$$\theta = t_I + T_C - T_A - T_B \tag{6-12}$$

This completes the second-order model.

Smith (5) and later Cox (6) used the graphical construction illustrated in Fig. 6-6 to determine the dead time. The steps are as follows:

1. Draw the line tangent to the inflection point as usual.
2. Let a denote the value of the response at the time at which the tangent line intersects the base line.
3. Locate point $b = 2.718a$, directly above point a.
4. Draw a line parallel to the original tangent line and passing through b.
5. The intersection of this line with the base line is the dead time.

For the response illustrated in Fig. 6-6, this method gives a dead time of 0.78. Using Eq. 6-11 gives a value of 0.77 via Sten's method.

The response in Figs. 6-4 and 6-6 is the step response of the overdamped process:

$$G(s) = \frac{1}{(0.5s + 1)\,(s + 1)^2\,(2s + 1)}$$

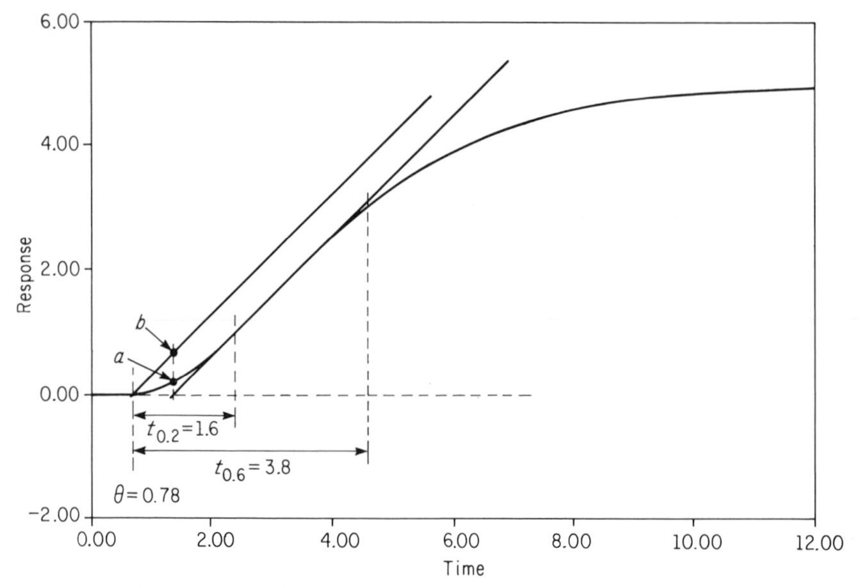

FIG. 6-6. Graphical construction to determine the dead time.

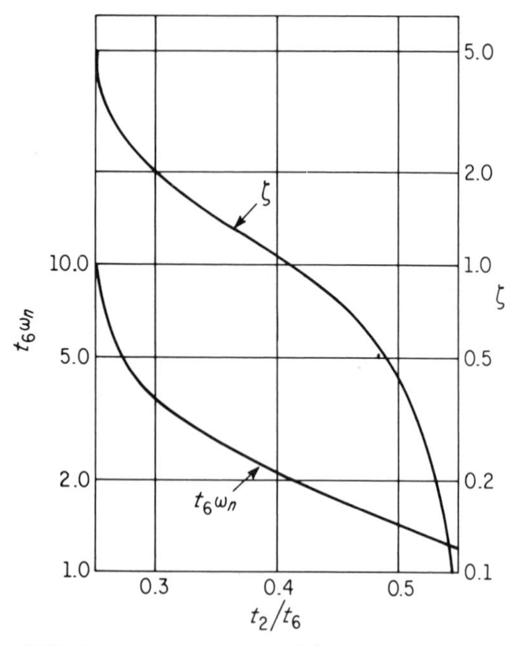

FIG. 6-7. Relationship of ζ and ω_n to t_2 and t_6. (Reproduced by permission from Ref. 7.)

The ratio of T_C/T_A for this response curve is about 0.66, which unfortunately does not give any solutions for τ/T_A using Fig. 6-5 or Eqs. 6-9 and 6-10. The problem is that values of T_C/T_A below 0.736 indicate that an underdamped model is in order.

Provided that an estimate of the dead time is available, Meyer et al. (7) presented a method by which the natural frequency and damping ratio can be determined. The dead time can be estimated using either the method proposed by Smith (5) or by Sten (3), although the latter does not exactly apply to underdamped cases. The required parameters to be ascertained from the response curve are t_2, the time past the dead time at which the response attains 20 percent of its final value, and t_6, the time past the dead time at which the response attains 60 percent of its final value. Thus the graph in Fig. 6-7 is used to determine values of ζ and ω_n, the values being 0.9 and 0.5 respectively for the response in Fig. 6-6. The quality of the approximation is illustrated in Fig. 6-8, and is substantially better than any of the first-order approximations in Fig. 6-3c.

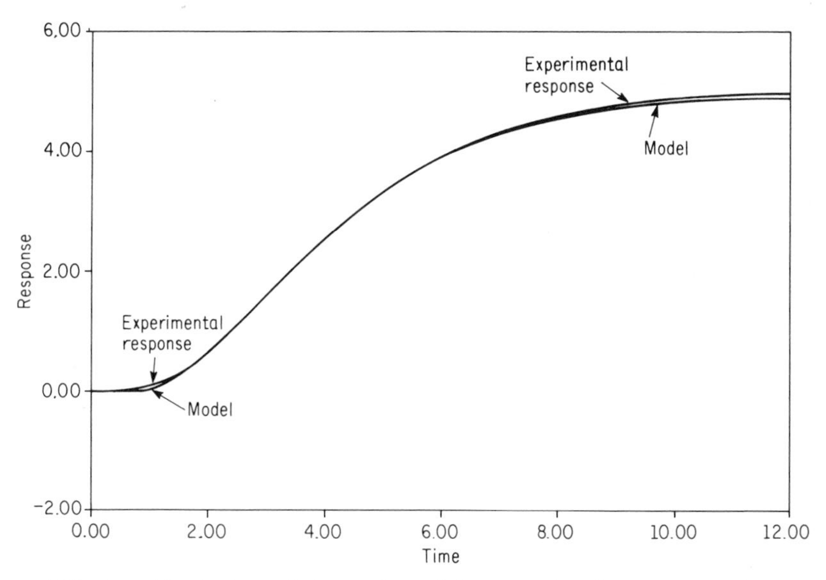

FIG. 6-8. Illustration of the quality of the second-order approximation.

6-2 DESIGN OF CONTROL ALGORITHMS USING z-TRANSFORMS

The general digital process control loop considered in this section is illustrated in Fig. 6-9. The objective is to design the controller $D(z)$ so that the desired loop performance is obtained. In Chapter 4 we saw that the block diagram in Fig. 6-9 can be represented by the fol-

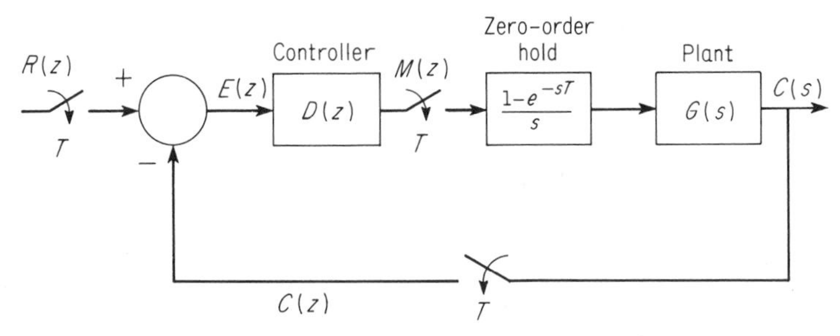

FIG. 6-9. Typical digital process control loop.

lowing equation:

$$C(z) = HG(z)D(z)\left[R(z) - C(z)\right] \tag{6-13}$$

where

$$HG(z) = \mathfrak{z}\left[\frac{1 - e^{-sT}}{s} \cdot G(s)\right] \tag{6-14}$$

is the pulse transfer function of the process.

In order to design $D(z)$, an expression for $HG(z)$ must be obtained in some manner. Once $HG(z)$ is specified, the only unknowns in Eq. 6-13 are $D(z)$ and $C(z)/R(z)$. If the desired loop-performance characteristics could be used to specify $C(z)/R(z)$, then Eq. 6-13 could be solved for $D(z)$:

$$D(z) = \frac{1}{HG(z)} \cdot \frac{C(z)/R(z)}{1 - C(z)/R(z)} \tag{6-15}$$

Therefore $D(z)$ can be readily calculated.

Before proceeding with specific examples, it should be pointed out that some attention must be given to $HG(z)$ when selecting $C(z)/R(z)$. For example, suppose the process contains a time delay equal to NT, i.e., $HG(z)$ contains the term z^{-N}. Then no matter what input to $HG(z)$ occurs, there will be no response for time NT. Therefore the ratio $C(z)/R(z)$ must also contain the time-delay term z^{-N}. If situations such as this are not recognized, the controller $D(z)$ would require future values of the error in order to calculate the current value of the output, which cannot physically be accomplished. Controllers of this type are said to be physically unrealizable.

One approach to obtaining expressions for $HG(z)$ is to use the shortcut modeling techniques described in the previous section.

For the approximations to the fourth-order system derived in the previous section, Table 6-2 gives numerical values for $HG(z)$ corresponding to three different sampling times. In subsequent sections,

TABLE 6-2
Model Pulse Transfer Functions

	First-Order Model	Second-Order Model
Continuous transfer function	$\dfrac{1.0e^{-1.46s}}{3.34s+1}$	$\dfrac{1.0e^{-0.78s}}{4s^2+3.6s+1}$
Pulse transfer function, $T = 5$	$\dfrac{z^{-1}(0.6535+0.1227z^{-1})}{1-0.2238z^{-1}}$	$\dfrac{z^{-1}(0.6634+0.2491z^{-1}+0.0011z^{-2})}{1-0.09754z^{-1}+0.01111z^{-2}}$
Pulse transfer function, $T = 1$	$\dfrac{z^{-2}(0.1493+0.1095z^{-1})}{1-0.7413z^{-1}}$	$\dfrac{z^{-1}(0.005664+0.1167z^{-1}+0.3910z^{-2})}{1-1.2451z^{-1}+0.4066z^{-2}}$
Pulse transfer function, $T = 0.2$	$\dfrac{z^{-8}(0.0410+0.0171z^{-1})}{1-0.9419z^{-1}}$	$\dfrac{z^{-5}(0.004709+0.004434z^{-1})}{1-1.8261z^{-1}+0.8353z^{-2}}$

controllers will be designed for the fourth-order process using these approximations.

Although we shall use these simple models throughout this chapter, the more elaborate modeling techniques described in Chapter 7 should produce a better model, from which a better control algorithm could be obtained. Just how much "better" is a matter of conjecture.

6-3 DEADBEAT ALGORITHMS (8)

The design procedure described above essentially reduces to specifying the outputs at the sampling instants for a particular test input. A deadbeat or minimal response is one that satisfies the criteria that (1) the settling time must be finite, (2) the rise time should be a minimum, and (3) the steady-state error should be zero.

One specific case that satisfies the above criteria is that the response to a step change in the set point should have zero error at all sampling instants after the first, as illustrated in Fig. 6-10. Specifically,

$$R(z) = \frac{1}{1 - z^{-1}} \quad \text{(a step input)}$$

$$\begin{aligned} C(z) &= z^{-1} + z^{-2} + \cdots \\ &= z^{-1}(1 + z^{-1} + z^{-2} + \cdots) \\ &= \frac{z^{-1}}{1 - z^{-1}} \end{aligned}$$

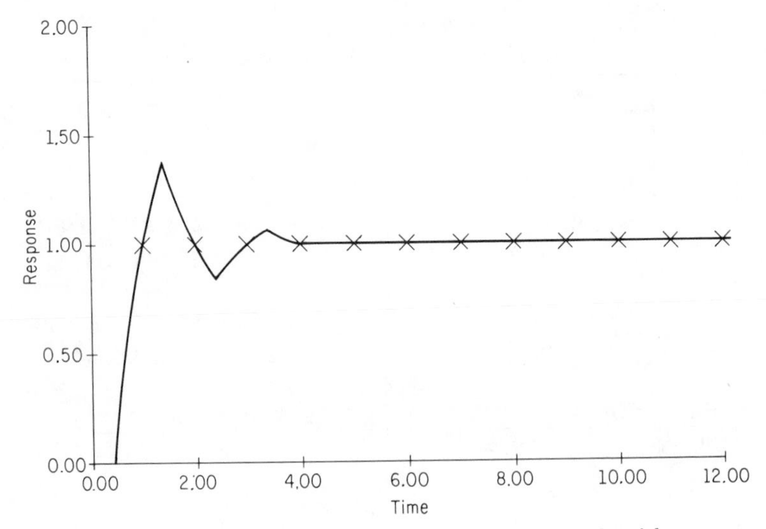

FIG. 6-10. Desired response using the deadbeat algorithm.

Therefore

$$\frac{C(z)}{R(z)} = z^{-1}$$

Substituting into Eq. 6-15 gives

$$D(z) = \frac{z^{-1}}{1 - z^{-1}} \cdot \frac{1}{HG(z)}$$

The control algorithm will be physically realizable only if the time delay in $HG(z)$ does not exceed the sampling time.

As a specific case, consider applying this design technique to the pulse transfer function $HG(z)$ in Table 6-2 corresponding to the first-order model and a sampling time of 5.0. The controller $D(z)$ would be

$$D(z) = \frac{1 - 0.2238z^{-1}}{z^{-1}(0.6535 + 0.1227z^{-1})} \cdot \frac{z^{-1}}{1 - z^{-1}}$$
$$= \frac{1 - 0.2238z^{-1}}{(0.6535 + 0.1227z^{-1})(1 - z^{-1})} = \frac{M(z)}{E(z)} \quad (6\text{-}16)$$

Note that this algorithm contains the term $(1 - z^{-1})$ in the denominator and therefore contains the integral mode.

In the time domain, the control algorithm is

$$m_n = (e_n - 0.2238e_{n-1} + 0.5308m_{n-1} + 0.1227m_{n-2})/0.6535 \quad (6\text{-}17)$$

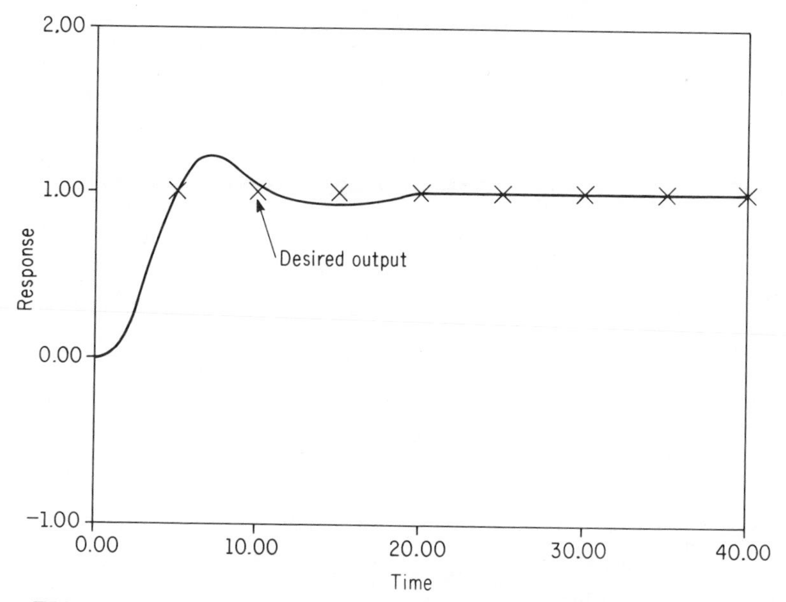

FIG. 6-11. Response of the fourth-order system using the control algorithm in Eq. 6-11.

The response of the original fourth-order system to a step change in set point is shown in Fig. 6-11. Due to the modeling errors, the response does not exactly equal 1.0 at the sampling instants.

If the sampling time is reduced to 1.0, the response criterion must be modified since the time delay now exceeds the sampling time. Although $c(0) = c(T) = 0$, we can require that all subsequent values equal 1. The output is now

$$C(z) = z^{-2} + z^{-3} + z^{-4} + \cdots = \frac{z^{-2}}{1 - z^{-1}}$$

and

$$\frac{C(z)}{R(z)} = z^{-2}$$

Therefore

$$\begin{aligned} D(z) &= \frac{1}{HG(z)} \cdot \frac{z^{-2}}{1 - z^{-2}} \\ &= \frac{1 - 0.7413z^{-1}}{(0.1493 + 0.1095z^{-1})(1 - z^{-2})} \end{aligned} \tag{6-18}$$

The performance of this algorithm, as illustrated in Fig. 6-12, is clearly unsatisfactory, for reasons that we shall explain shortly.

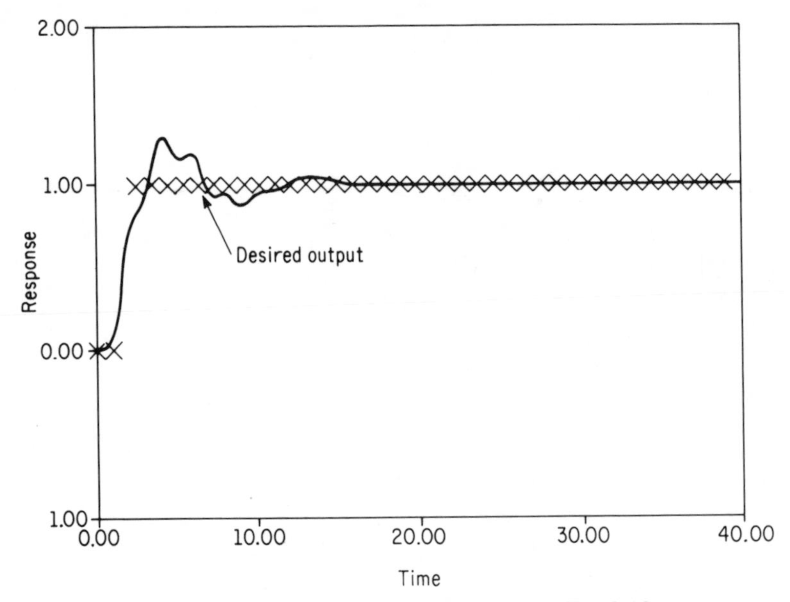

FIG. 6-12. Performance of algorithm in Eq. 6-18.

6-4 DAHLIN'S METHOD

Requiring the process output to move from the old set point to the new set point in one sampling time is far too strenuous for most industrial processes. The problem is not with the design method, but with the chosen expression for $C(z)/R(z)$. Instead of the deadbeat, Dahlin (9) suggests that the closed-loop system should behave like a continuous first-order lag with dead time. For the continuous loop, the response to a step change in set point would be

$$C(z) = \frac{e^{-\theta s}}{(\lambda s + 1)} \cdot \frac{1}{s} \tag{6-19}$$

where λ is the time constant of the closed-loop response. In discrete form, this equation is

$$C(z) = \frac{(1 - e^{T/\lambda})\, z^{-N-1}}{(1 - z^{-1})\,(1 - e^{T/\lambda} z^{-1})} \tag{6-20}$$

where N = nearest integer number of sampling times in θ.

If $R(z)$ is the unit step, then

$$\frac{C(z)}{R(z)} = \frac{(1 - e^{T/\lambda}) z^{-N-1}}{1 - e^{-T/\lambda} z^{-1}} \tag{6-21}$$

The control algorithm $D(z)$ can be readily calculated.

$$D(z) = \frac{C(z)/R(z)}{1 - C(z)/R(z)} \cdot \frac{1}{HG(z)}$$

$$= \frac{(1 - e^{-T/\lambda})z^{-N-1}}{1 - e^{-T/\lambda}z^{-1} - (1 - e^{-T/\lambda})z^{-N-1}} \cdot \frac{1}{HG(z)} \tag{6-22}$$

The time constant λ of the closed-loop response can be best visualized as a tuning parameter. Small values of λ give tight control; large values of λ give loose control.

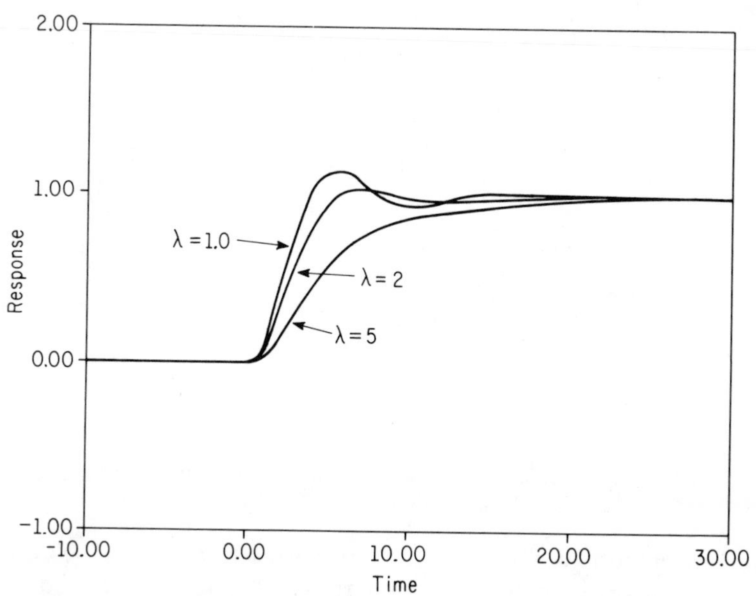

FIG. 6-13. Performance of Dahlin's algorithm for various values of λ.

Figure 6-13 illustrates the effect of λ on the responses of the fourth-order system using the $HG(z)$ corresponding to the first-order approximation and a sampling time of 1.0. The response corresponding to $\lambda = 2.0$ would probably be preferred by most designers. The control algorithm is

$$D(z) = \frac{0.292(1 - 0.741z^{-1})}{(1 - 0.608z^{-1} - 0.392z^{-2})(0.149 + 0.110z^{-1})}$$

$$= \frac{1.96(1 - 0.741z^{-1})}{(1 - z^{-1})(1 + 0.392z^{-1})(1 + 0.738z^{-1})} \tag{6-23}$$

$$= \frac{1.96z^2(z - 0.741)}{(z - 1)(z + 0.392)(z + 0.738)}$$

We shall refer to this algorithm again in the next section.

6-5 RINGING

Judging the performance of a control algorithm solely from the response $c(t)$ is generally not sufficient. For example, the response of Dahlin's algorithm for $\lambda = 2.0$ as shown in Fig. 6-13 looks quite good. However, the valve signal $m(t)$ shown in Fig. 6-14 exhibits excessive

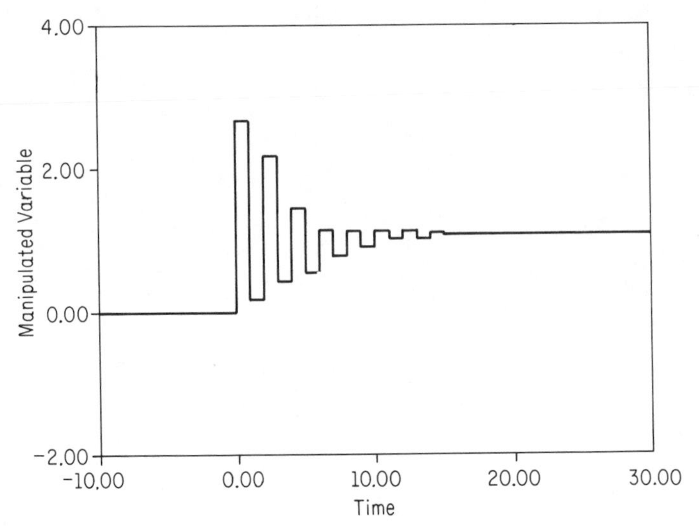

FIG. 6-14. Valve action for response in Fig. 6-13 corresponding to $\lambda = 2$.

valve movements, a phenomenon known as *ringing* (9). This may exist with algorithms designed by virtually any technique, and is certainly not unique to Dahlin's algorithms.

The effect of pole-zero locations on ringing can be appreciated from an examination of Fig. 6-15. The $z = -1$ point is called the *ringing node*, and the pole at $z = -1$ is known as the *ringing pole*. From the first entry in Fig. 6-15, the ringing amplitude, the difference between the first two controller outputs, of this element is 1.0. Moving the pole away from the ringing node decreases the ringing amplitude, as evidenced by the second entry ($z = -0.5$) in Fig. 6-15.

As illustrated by the third entry in Fig. 6-15, a pole in the right-half plane reduces the ringing amplitude, while zeros in the right-half plane aggravate it.

As an example, Fig. 6-16 shows the pole-zero locations in the z-plane for the algorithm in Eq. 6-23. The two poles located at $z = -0.738$ and -0.392 would lead one to suspect this algorithm of ringing, which is certainly verified by Fig. 6-14. The question is what to do about it.

	$D(Z)$	Impulse Response	Step Response	RA	Time Plots
1	$\frac{1}{1+1/z}$	1 −1 1 −1 1	1 0 1 0 1	1	1, D(z) OUTPUT, RA, 0, TIME
2	$\frac{1}{1+.5/z}$	1 −.5 .25 −.125	1 .5 .75 .625	.5	1, D(z) OUTPUT, RA, SETTLING VALUE, 0, TIME
3	$\frac{1}{(1+.5/z)(1-.2/z)}$	1 −.3 .19 −.087 .045	1 .7 .89 .803 .848	.3	1, D(z) OUTPUT, RA, SETTLING VALUE, 0, TIME
4	$\frac{1-.5/z}{(1+.5/z)(1-.2/z)}$	1 −.8 .36 −.188 .0924	1 .2 .50 .37 .46	.8	1, D(z) OUTPUT, RA, SETTLING VALUE, 0, TIME

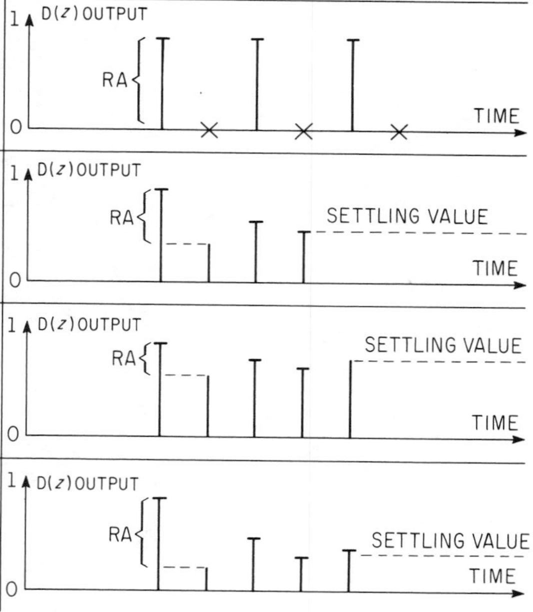

FIG. 6-15. Ringing characteristics of representative systems. (Reprinted by permission from Ref. 9.)

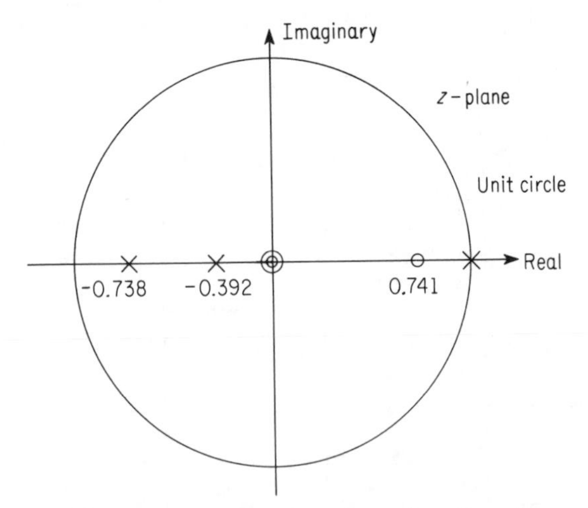

FIG. 6-16. Pole-zero locations for the algorithm in Eq. 6-23.

For the most part, the ringing phenomenon is generally superimposed upon the useful output of the algorithm. Dahlin (9) suggests that the ringing pole be simply eliminated and the gain adjusted accordingly. Since the steady state gain of any term is found by letting z approach one, this is done by simply letting $z = 1$ in the term to be eliminated. Replacing the term $1 + 0.738z^{-1}$ in Eq. 6-23 by $1 + 0.738 = 1.738$ gives the algorithm

$$D(z) = \frac{1.13(1 - 0.741z^{-1})}{(1 - z^{-1})(1 + 0.392z^{-1})} \tag{6-24}$$

The response $c(t)$ and the corresponding valve action shown in Fig. 6-17 show that the algorithm is now essentially ringing-free. The response $c(t)$ is very similar to that in Fig. 6-13 for the ringing algorithm.

The deadbeat algorithm in Eq. 6-18 is also subject to ringing. Factoring the denominator gives

$$D(z) = \frac{6.7(1 - 0.7413z^{-1})}{(1 + 0.733z^{-1})(1 - z^{-1})(1 + z^{-1})}$$

With poles at -1 and -0.733, this algorithm rings considerably.

6-6 KALMAN'S APPROACH

Instead of specifying $C(z)/R(z)$, Kalman placed restrictions on $M(z)$ and $C(z)$ in order to obtain an algorithm (10). We shall illustrate

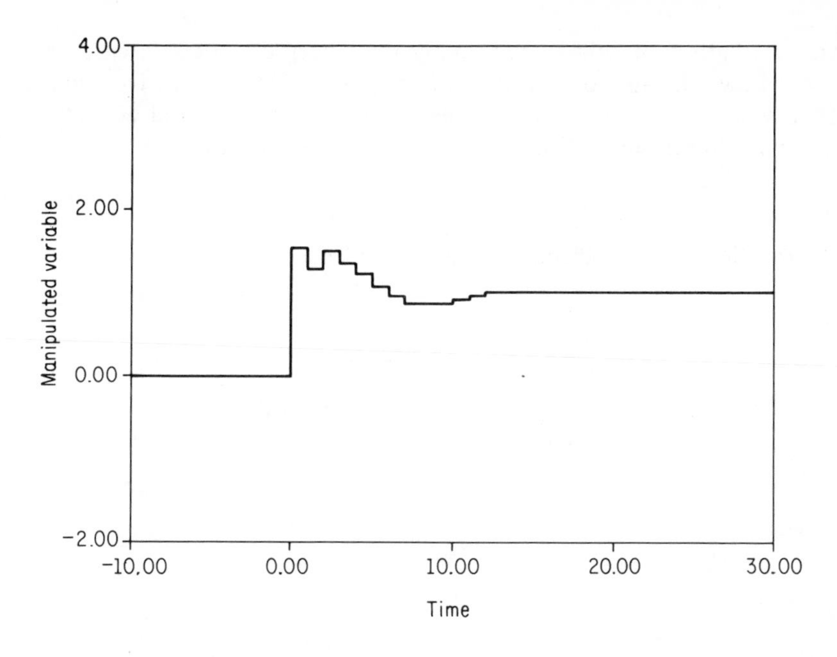

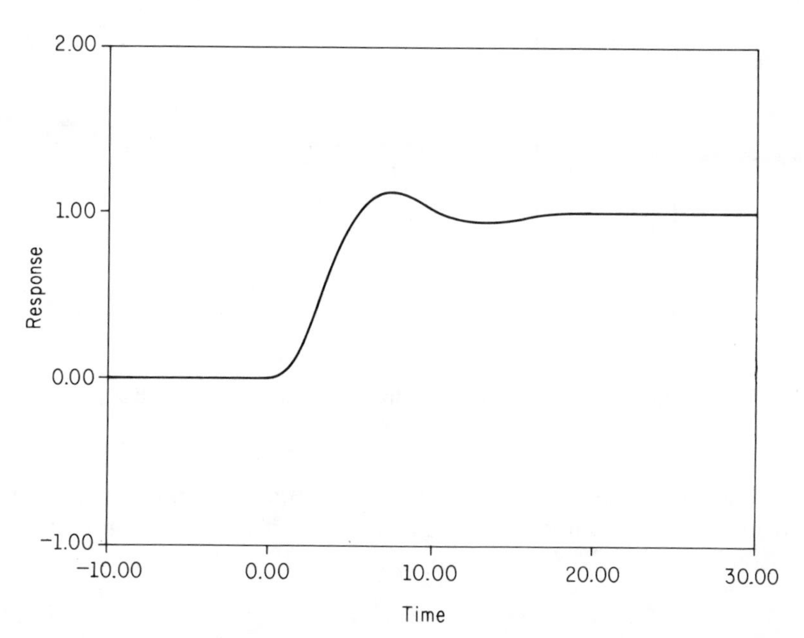

FIG. 6-17. Response and value action for Dahlin's algorithm with ringing pole removed ($\lambda = 2$, $T = 1$.)

the derivation of this algorithm by considering a specific case. Suppose we would like our system's response to a step input to reach the final value in two sampling times and remain at the final value thereafter, as illustrated in Fig. 6-18. The expression for $C(z)$ is

$$C(z) = c_1 z^{-1} + z^{-2} + z^{-3} + \cdots \tag{6-25}$$

No restrictions are placed on the value of c_1.

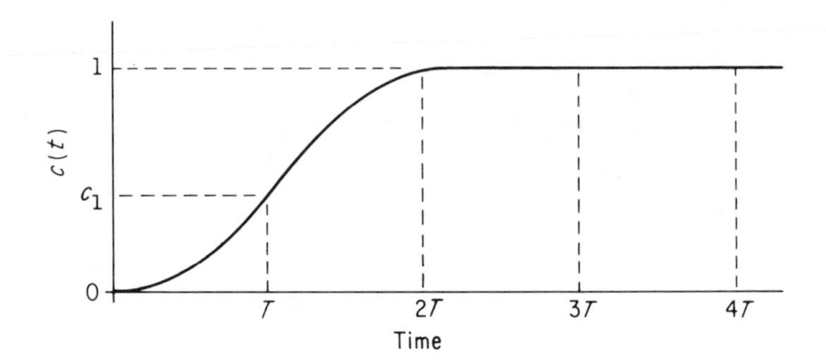

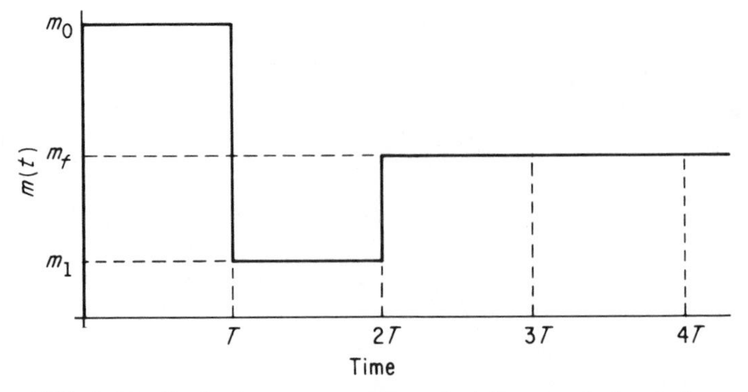

FIG. 6-18. Desired response characteristics used to design Kalman's algorithm.

In order to accomplish this, the manipulated variable will assume two intermediate values and then assume its final value thereafter, as illustrated in Fig. 6-18. The expression for $M(z)$ is

$$M(z) = m_0 + m_1 z^{-1} + m_f z^{-2} + m_f z^{-3} + \cdots \tag{6-26}$$

Note that the value of m_f equals the reciprocal of the process gain.

Since the input is a unit step change in the set point, $R(z)$ equals $1/(1 - z^{-1})$.

$$\frac{C(z)}{R(z)} = (1 - z^{-1})(c_1 z^{-1} + z^{-2} + z^{-3} + \ldots)$$

$$= c_1 z^{-1} + (1 - c_1) z^{-2}$$

$$= p_1 z^{-1} + p_2 z^{-1} = P(z) \qquad (6\text{-}27)$$

$$\frac{M(z)}{R(z)} = (1 - z^{-1})(m_0 + m_1 z^{-1} + m_f z^{-2} + m_f z^{-3} + \cdots)$$

$$= m_0 + (m_1 - m_0)z^{-1} + (m_f - m_1)z^{-2}$$

$$= q_0 + q_1 z^{-1} + q_2 z^{-2} = Q(z) \qquad (6\text{-}28)$$

Recall that the process pulse transfer function $HG(z)$ is the ratio of $C(z)$ to $M(z)$.

$$HG(z) = \frac{C(z)}{M(z)} = \frac{P(z)}{Q(z)} \qquad (6\text{-}29)$$

Therefore we see that the coefficients in $P(z)$ and $Q(z)$ must equal the coefficients in the process pulse transfer function $HG(z)$.

The following relationships among the coefficients should be noted.

$$\sum_{i=1}^{2} p_i = p_1 + p_2 = 1 \qquad (6\text{-}30)$$

$$\sum_{i=0}^{2} q_i = q_0 + q_1 + q_2 = m_f = 1/K \qquad (6\text{-}31)$$

where K is the process gain. These relationships do not generally hold for the pulse transfer function as typically derived. But simply dividing by the sum of the numerator coefficients will insure that both hold.

Now that $P(z)$ and $Q(z)$ are known, the control algorithm $D(z)$ can be derived in the usual manner:

$$D(z) = \frac{1}{HG(z)} \cdot \frac{C(z)/R(z)}{1 - C(z)/R(z)}$$

$$= \frac{Q(z)}{P(z)} \cdot \frac{P(z)}{1 - P(z)} = \frac{Q(z)}{1 - P(z)} \qquad (6\text{-}32)$$

Therefore the coefficients in the control algorithm are directly related to the coefficients in the process pulse transfer function.

Note that we assumed two intermediate switches in the manipulated variable and found that the denominator of $HG(z)$ must contain terms through z^{-2}, or in other words, the plant should be second order. The number of intermediate switches is indeed equal to the order of the process. That is, a third-order process would require three switches.

If the process contains a dead time term z^{-N}, Eq. 6-32 still holds. That is, process dead time presents no problems in deriving the control algorithm.

As an example, consider deriving the control algorithm corresponding to the entry in Table 6-2 for the first-order model and a sampling time of 1.0:

$$HG(z) = \frac{z^{-2}(0.1493 + 0.1095z^{-1})}{1 - 0.7413z^{-1}}$$

We begin by normalizing the numerator.

$$HG(z) = \frac{z^{-2}(0.577 + 0.423z^{-1})}{3.86 - 2.86z^{-1}} = \frac{P(z)}{Q(z)} \tag{6-33}$$

The control algorithm is then

$$D(z) = \frac{3.86 - 2.86z^{-1}}{1 - z^{-2}(0.577 + 0.423z^{-1})} \tag{6-34}$$

The performance of this algorithm is illustrated in Fig. 6-19.

6-7 DISCRETE EQUIVALENT TO AN ANALOG CONTROLLER

While the design techniques in the previous sections have relied on z-transforms, another route is to design an analog controller and then use its discrete counterpart (9). For example, suppose the transfer function $G(s)$ is a first-order-lag-plus-dead-time system.

$$G(s) = \frac{Ke^{-\theta s}}{\tau s + 1}$$

Substituting into the continuous equivalent of Eq. 6-15,

$$D(s) = \frac{C(s)}{R(s)} = \frac{\tau s + 1}{Ke^{-\theta s}} \cdot \frac{C(s)/R(s)}{1 - C(s)/R(s)} \tag{6-35}$$

The selection of $C(s)/R(s)$ is quite analogous to the selection of $C(z)/M(z)$ in the previous sections. As for the discrete case, to be physically realizable $D(s)$ must not contain any predictive elements (negative dead times). By inspection of Eq. 6-35 this means that $C(s)/R(s)$ must also contain the dead time $e^{-\theta s}$.

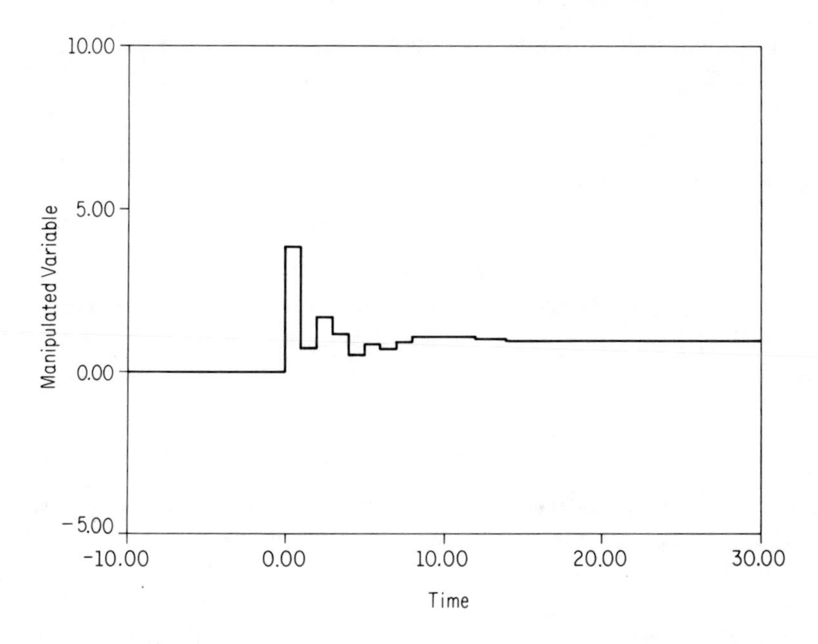

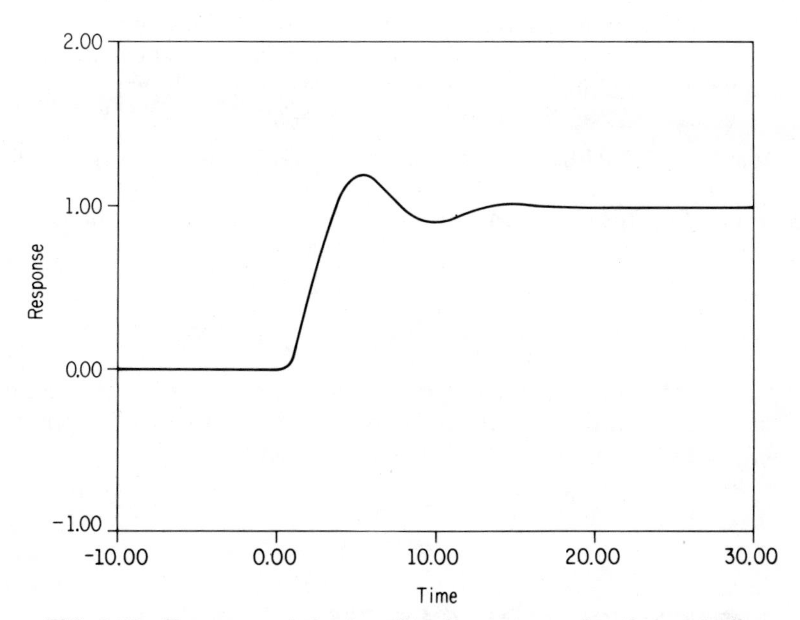

FIG. 6-19. Response of the fourth-order system using the controller in Eq. 6-34.

As an example, suppose we choose $C(s)/R(s)$ to be

$$\frac{C(s)}{R(s)} = \frac{e^{-\theta s}}{\lambda s + 1}$$

Equation 6-35 then gives the following expression for $D(s)$:

$$D(s) = \frac{(\tau s + 1)/K}{\lambda s + (1 - e^{-\theta s})}$$

The next step is to determine a discrete equivalent to $D(s)$. If $D(s) = M(s)/E(s)$, then in the time domain the differential equation describing $D(s)$ is

$$\lambda \frac{dx(t)}{dt} + x(t) - x(t - \theta) = \left[\tau \frac{de(t)}{dt} + e(t)\right] /K$$

Expressing the derivatives in terms of forward differences and writing the discrete approximation to this equation at the $(n - 1)$st sampling time gives

$$\lambda \frac{m_n - m_{n-1}}{T} + m_{n-1} - m_{n-N-1} = \left[\tau \frac{e_n - e_{n-1}}{T} + e_{n-1}\right] /K$$

where T is the sampling time and N is the nearest integer number of sampling times in θ. This equation can be solved for the value of the manipulated variable at the next sampling time as follows:

$$m_n = \left(1 - \frac{T}{\lambda}\right) m_{n-1} + \frac{T}{\lambda} m_{n-N-1} + \frac{\tau}{K\lambda} \left[e_n + \left(\frac{T}{\tau} - 1\right) e_{n-1}\right] \quad (6\text{-}36)$$

Alternatively, this equation can be expressed in the z-domain as

$$D(z) = \frac{\frac{\tau}{K\lambda}\left[1 - \left(\frac{T}{\tau} - 1\right) z^{-1}\right]}{1 - (1 - T/\lambda) z^{-1} - (T/\lambda) z^{-N-1}}$$

As one of the factors contributing heavily to the validity of these expressions is the finite difference approximations in Eq. 6-36, it is suggested that these equations only be used when

$$T \leqslant 0.2\tau$$

$$\lambda \geqslant 2.5T$$

6-8 DESIGN FOR LOAD CHANGES

The more general control loop encountered in process systems is shown in Fig. 6-20. This loop differs from the loops previously con-

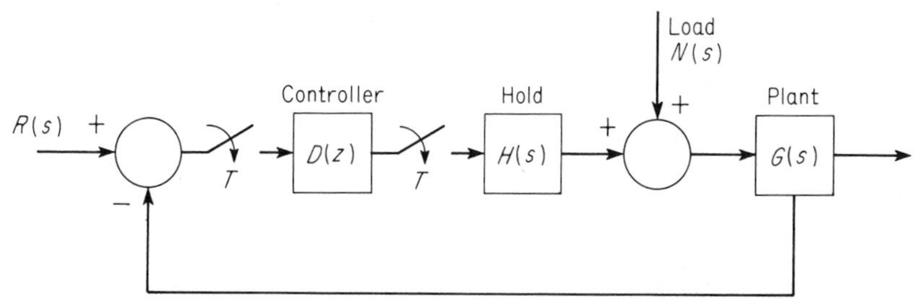

FIG. 6-20. Control loop subject to load changes.

sidered in that a disturbance $N(s)$ enters the loop in addition to a set point $R(s)$. The techniques discussed in the previous sections were for designing control algorithms for set-point changes. Therefore the question arises as to how they will perform for changes in the disturbance.

Fortunately, in many cases algorithms designed by these techniques perform very well for load changes also. For example, Figs. 6-21a and 6-21b illustrate the performance of Dahlin's algorithm and Kalman's algorithm for load changes. However, if the designer was mainly interested in load response, he would probably prefer to design specifically for this case rather than design for the set-point case and hope his algorithm works for the load change. This is the subject of the present section.

Applying the block diagram manipulations with z-transforms as discussed earlier for the system in Fig. 6-20 with $R(s)$ set equal to zero yields

$$C(z) = \frac{NG(z)}{1 + D(z)\,HG(z)}$$

As before, this equation can be solved for $D(z)$ in terms of the other transfer functions:

$$D(z) = \frac{NG(z) - C(z)}{HG(z)\,C(z)} \tag{6-37}$$

Superficially, the design procedure is very similar to the set-point case. The steps are:

1. Select an input $N(s)$.
2. Determine the desired response $C(s)$.
3. Take the transforms indicated into Eq. 6-37.
4. Solve for $D(z)$.

However, a few points of precaution are in order.

In the set-point case, the change could only enter the loop at the sampling instant. But in the disturbance case, the point in time

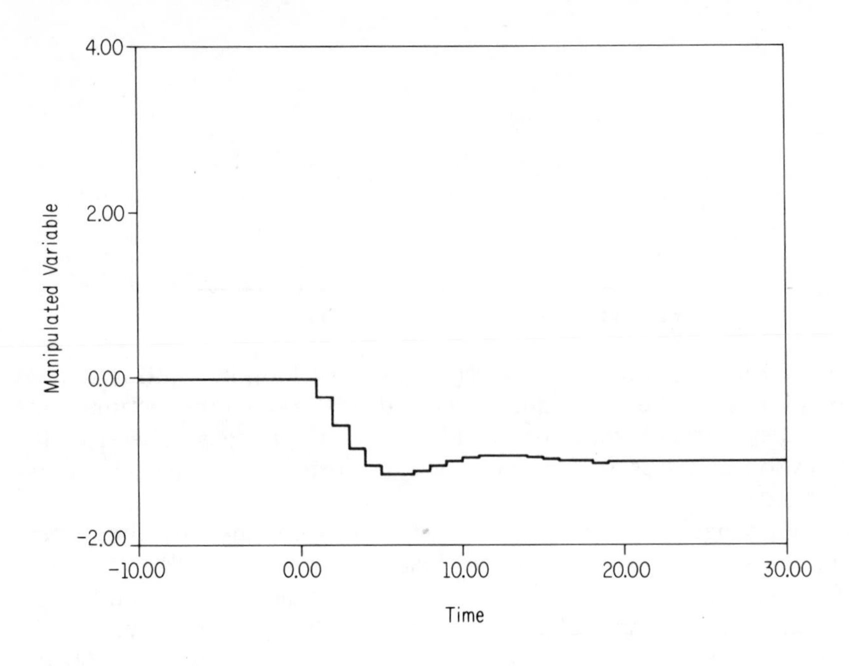

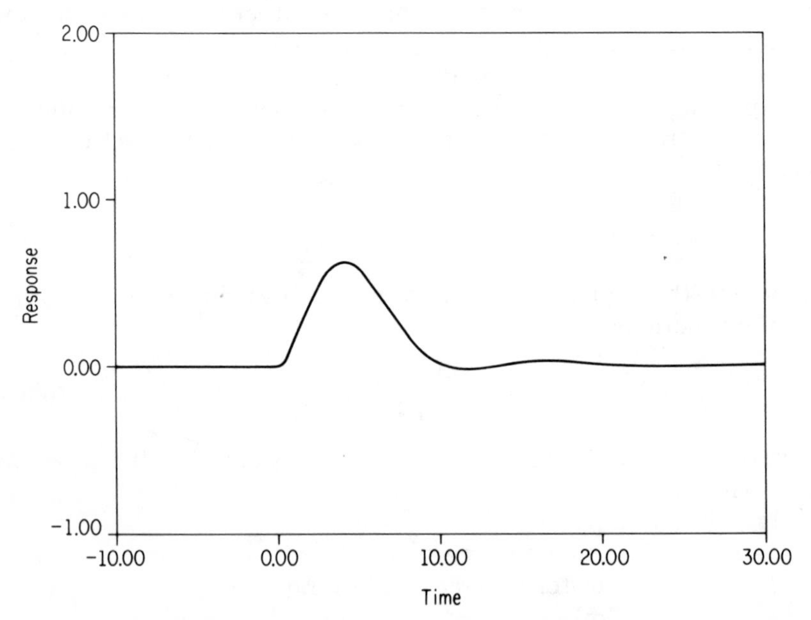

FIG. 6-21a. Response of Dahlin's algorithm (Eq. 6-23) to a step change in load ($\lambda = 2.0$, $T = 1.0$).

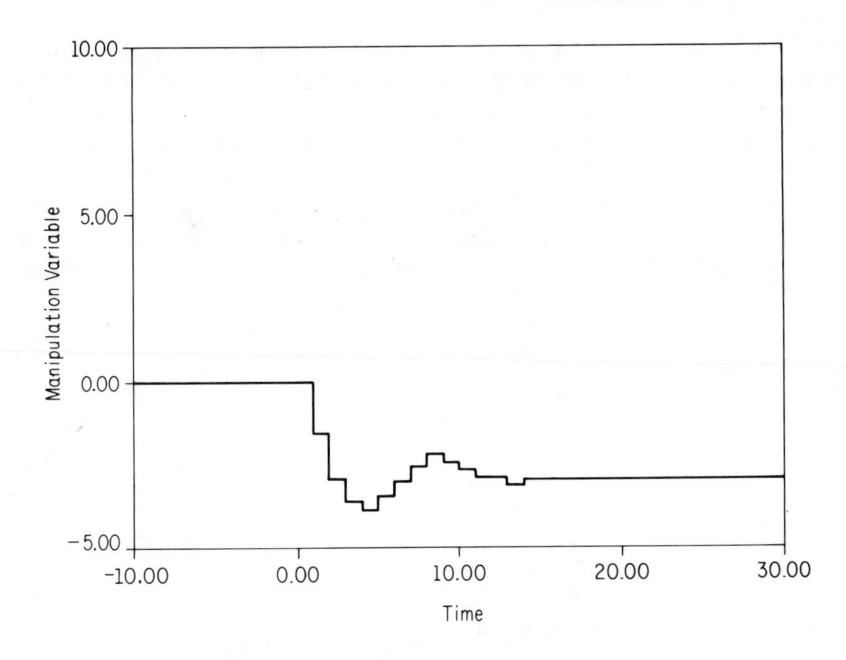

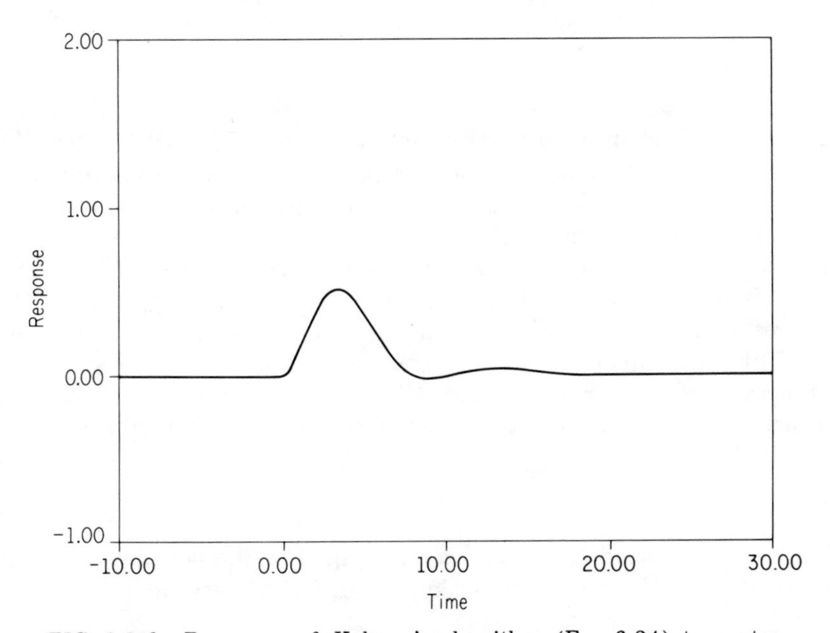

FIG. 6-21b. Response of Kalman's algorithm (Eq. 6-34) to a step change in load ($T = 1$).

relative to the sampling instant that the disturbance enters the loop is not normally under the designer's control. Although this affects the performance of the loop, it has been reported (11) to be rather unimportant in most cases. In such cases the common engineering approach is to design for the worst possible case. If the dead time of the process is θ, then the worst possible instant at which to admit the disturbance is θ units of time prior to the sampling instant. Refer to Fig. 6-22. Note that a unit step disturbance enters θ units of time

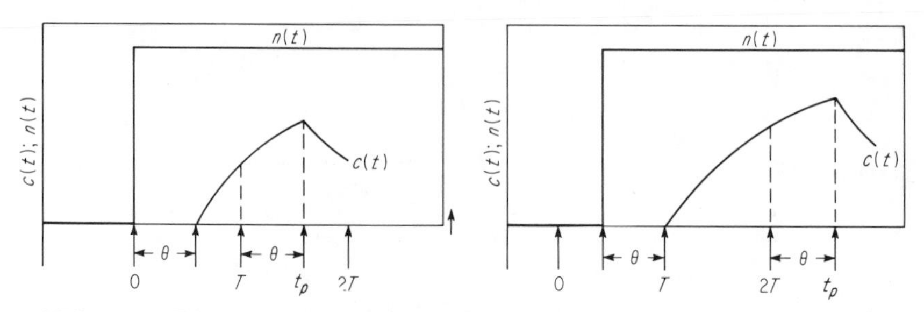

FIG. 6-22. Effect on the system response of time at which a disturbance signal enters a loop relative to a sampling instant (left) signal introduced at sampling instant (right) disturbance introduced θ units before a sampling instant. (Reprinted by permission from Ref. 15.)

prior to the sampling instant T. As can be seen, the controller is not aware of this change until the sampling instant $2T$, at which time the controller would presumably prescribe a change in the manipulated variable to offset the disturbance. But this change is not reflected in the output until an additional θ units of time. The net effect is that the system is effectively not under control for time $\theta + T$ after the disturbance enters the loop, and no reaction appears in the process output until time $2\theta + T$ after the disturbance enters the loop.

This phenomenon also places restrictions on the allowable choices of $C(z)$ for a selected $N(s)$. To illustrate these, consider design of a controller for the first-order-lag-plus-dead-time process:

$$G(s) = \frac{e^{-Ts}}{\tau s + 1}$$

Note that the dead time has been set equal to the sampling time for convenience. If a zero-order hold is used, $HG(z)$ becomes

$$HG(z) = \frac{(1 - b)z^{-2}}{1 - bz^{-1}}$$

where $b = e^{-T/\tau}$.

If $N(s)$ is a unit step function admitted at time zero, then $NG(z)$ is

$$NG(z) = \mathfrak{Z}\left[\frac{e^{-Ts}}{s(\tau s + 1)}\right]$$

$$= \frac{(1 - b)}{(z - 1)(z - b)} = \frac{(1 - b)z^{-2}}{(1 - z^{-1})(1 - bz^{-1})}$$

This specifies all terms required in Eq. 6-37 except $C(z)$.

A typical output over the first three sampling instants for this input might appear as in Fig. 6-23. As the dead time equals one

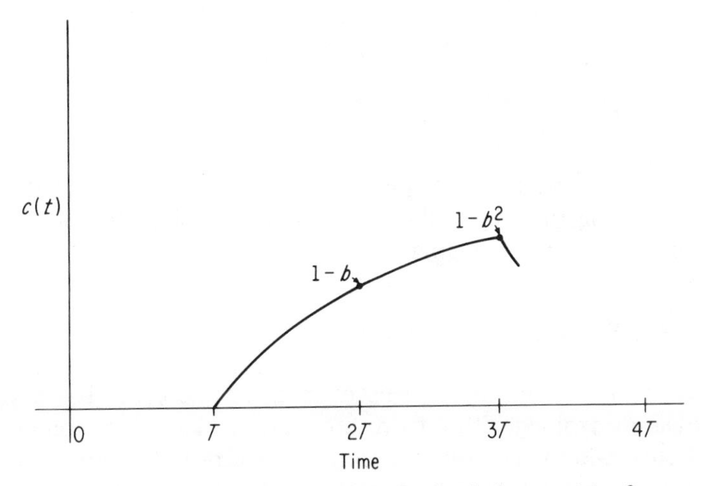

FIG. 6-23. Response to a unit step load change at $t = 0$.

sampling time, the output for the first two sampling instants must be zero. That is,

$$c(0) = 0$$

$$c(T) = 0$$

As no control action is taken at T, the response is the open-loop step response over the interval $T \leqslant t \leqslant 2T$. Thus,

$$c(2T) = 1 - e^{-T/\tau} = 1 - b$$

Although a control action is taken at $2T$, it is not reflected in the output until $3T$. The response is therefore effectively open loop over the interval $2T \leqslant t \leqslant 3T$, giving

$$c(3T) = 1 - e^{-2T/\tau} = 1 - b^2$$

From this point on we have rather complete freedom in selecting the response. Thus we might require that the response at subsequent sampling intervals be zero, or

$$c(nT) = 0, \qquad n > 3$$

From the definition of the z-transform, we can evaluate $C(z)$ as being

$$C(z) = \sum_{i=0}^{\infty} c(iT) z^{-i}$$

$$= (1 - b) z^{-2} + (1 - b^2) z^{-3}$$

Substituting this and the other transforms into Eq. 6-37,

$$D(z) = \frac{1 + b + b^2}{1 - b} \cdot \frac{\left[1 - \dfrac{b(1 + b) z^{-1}}{1 + b + b^2}\right]}{(1 - z^{-1})\,[1 + (1 + b) z^{-1}]} \tag{6-38}$$

It should be noted that the pole at $z = -(1 + b)$ is outside the unit circle, rendering this controller open loop unstable. This pole would also cause considerable ringing.

6-9 USE OF STANDARD ALGORITHMS

The previous sections of this chapter have treated the development of a control algorithm from the plant transfer function and the desired closed-loop transfer function. The alternative to this avenue parallels the common practice with analog controllers. That is, a given form of the algorithm is used and the desired response obtained as nearly as possible by adjusting the parameters, i.e., tuning the algorithm.

Perhaps the most common algorithm in use today is the discrete equivalent of the proportional-plus-integral-plus-derivative (PID) controller or its simplified cousin, the PI algorithm. This algorithm is derived by simply replacing the integral by a sum and the derivative by a finite difference. Using rectangular integration, this algorithm is

$$m_n = K_c \left[e_n + \frac{T}{T_i} \sum_{k=0}^{n} e_k + \frac{T_d}{T} (e_n - e_{n-1}) \right] + M_R \tag{6-39}$$

where K_c = proportional gain
T_i = reset time
T_d = derivative time
M_R = initial valve position

This is known as the *position form* of the algorithm, since the actual valve position is calculated from the error sequence.

An alternative is to calculate the change in valve (actuator) position rather than its actual value. Writing Eq. 6-39 for m_{n-1} gives

$$m_{n-1} = K_c \left[e_{n-1} + \frac{T}{T_i} \sum_{k=0}^{n-1} e_k + \frac{T_d}{T} (e_{n-1} - e_{n-2}) \right] + M_R .$$

Subtracting from Eq. 6-39 gives the velocity form of the algorithm:

$$\begin{aligned} m_n - m_{n-1} &= \Delta m_n \\ &= K_c \left[(e_n - e_{n-1}) + \frac{T}{T_i} e_n + \frac{T_d}{T} (e_n - 2e_{n-1} + e_{n-2}) \right] \end{aligned} \tag{6-40}$$

The advantage of this algorithm is that it does not have to be initialized. For example, if the position algorithm in Eq. 6-39 is initialized with the error and the sum equal to zero, the calculated value of the output is M_R. In practice, the operator generally will have the actuator set at some desired value before switching to automatic, and it would be desirable that the algorithm be initialized to maintain this position until an error is encountered. This can be accomplished by simply setting M_R equal to the valve position at the time control is transferred to the computer. But the computer must have available a numerical value for the actuator position, which must either be entered by the operator (a rather bothersome chore) or by an analog input (which costs money and burdens the analog front end). The simplest alternative is to use the velocity form of the algorithm which does not need to be initialized.

A second advantage, namely some protection against reset windup, of the velocity algorithm will be discussed shortly.

Another apparent consideration in the PID algorithm is the numerical integration technique for the integral. Using the trapezoid rule, Eq. 6-39 becomes

$$m_n = K_c' \left[e_n + \frac{T}{T_i'} \sum_{k=0}^{n} \frac{e_k + e_{k-1}}{2} + \frac{T_d'}{T} (e_n - e_{n-1}) \right]$$

Expressing in the velocity form,

$$\Delta m_n = K_c' \left[e_n - e_{n-1} + \frac{T}{2T_i'} (e_n + e_{n-1}) + \frac{T_d'}{T} (e_n - 2e_{n-1} + e_{n-2}) \right]$$

or equivalently,

$$\Delta m_n = K_c' \left[\left(1 + \frac{T}{2T_i'} + \frac{T_d'}{T}\right) e_n + \left(\frac{T}{2T_i'} - 2\frac{T_d'}{T} - 1\right) e_{n-1} + \frac{T_d'}{T} e_{n-2} \right]$$

Collecting terms in Eq. 6-40 (the velocity algorithm developed using rectangular integration) gives

$$\Delta m_n = K_c \left[\left(1 + \frac{T}{T_i} + \frac{T_d}{T}\right) e_n - \left(1 + 2\frac{T_d}{T}\right) e_{n-1} + \frac{T_d}{T} e_{n-2} \right]$$

Equating terms gives the following relationships among the parameters:

$$K_c = K_c' \left(1 - \frac{T}{2T_i'}\right)$$

$$T_i = T_i' - T/2$$

$$T_d = \frac{2T_d' T_i'}{2T_i' - T}$$

Thus the algorithm parameters are directly related, and if the proper values are used, the control provided by these two will be identical.

Several modifications of the above algorithms are in common use. In most cases there is no real advantage in allowing the derivative mode to act on a change in the set point signal. In the computer the feedback variable b_n can be just as easily used as the error sequence e_n. In these cases Eq. 6-39 becomes

$$m_n = K_c \left[e_n + \frac{T}{T_i} \sum_{i=0}^{n} e_i - \frac{T_d}{T} (b_n - b_{n-1}) \right] + M_R$$

The velocity form of the algorithm is analogous.

Another common feature is the insertion of an "antiwindup" feature for the integral mode (12). With the position form of the algorithm, a couple of possibilities are available. One of these is to simply place an upper limit on the sum. A second possibility is to detect when the control element saturates and terminate the summation until the manipulated variable returns to within the control range. It should perhaps be noted that the last option is automatically incorporated into the velocity algorithm.

Another possible modification of the algorithm will be discussed in a later section.

6-10 TUNING THE ALGORITHM

The critical phase of the implementation of the PI or PID algorithm is the selection of the numerical values of the constants in the algorithm. The steps in tuning a digital control algorithm parallel the steps in tuning an analog controller. These are

1. Approximate the process with a simple model.
2. Select the constants that give the desired behavior when controlling the model.
3. Apply these settings to the original process.

For controller tuning, the first-order-lag-plus-dead-time model as derived in Sec. 6-1 is almost invariably used, although a couple of approaches using second-order-lag-plus-dead-time models are available (13,14).

For a PI controller, the control loop with the first-order-lag-plus-dead-time model is shown in Fig. 6-24. For a selected input (for

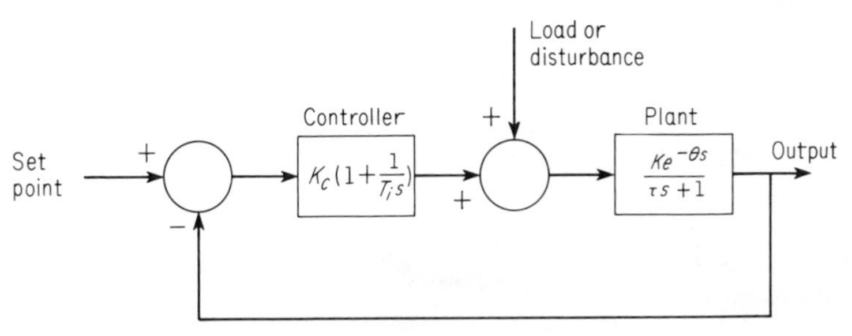

FIG. 6-24. Control loop with PI controller and first-order-lag-plus-dead-time system.

example, step change in set point) the response is a function of only the controller parameters, i.e., K_c and T_i. The objective of the tuning process is to select the combination of K_c and T_i that gives the response $c(t)$ that most nearly satisfies the desired criteria. This is expressed as

$$\underset{K_c, T_i}{\text{optimize}} \{c(t)\}$$

which reads "optimize the response $c(t)$ over K_c and T_i."

The criteria upon which to base the selection of the "best" $c(t)$ vary from application to application, system to system, and individual to individual. The commonly used criteria can be separated into two classes: (1) the simple but approximate criteria based on only a few points on the response, and (2) the more exact (but more difficult to evaluate) criteria based on the entire response, (e.g., integral criteria).

In the first category fall such criteria as settling time, percent overshoot, rise time, decay ratio, and the like which are defined in Fig. 6-25. Of these, the decay ratio has been commonly accepted

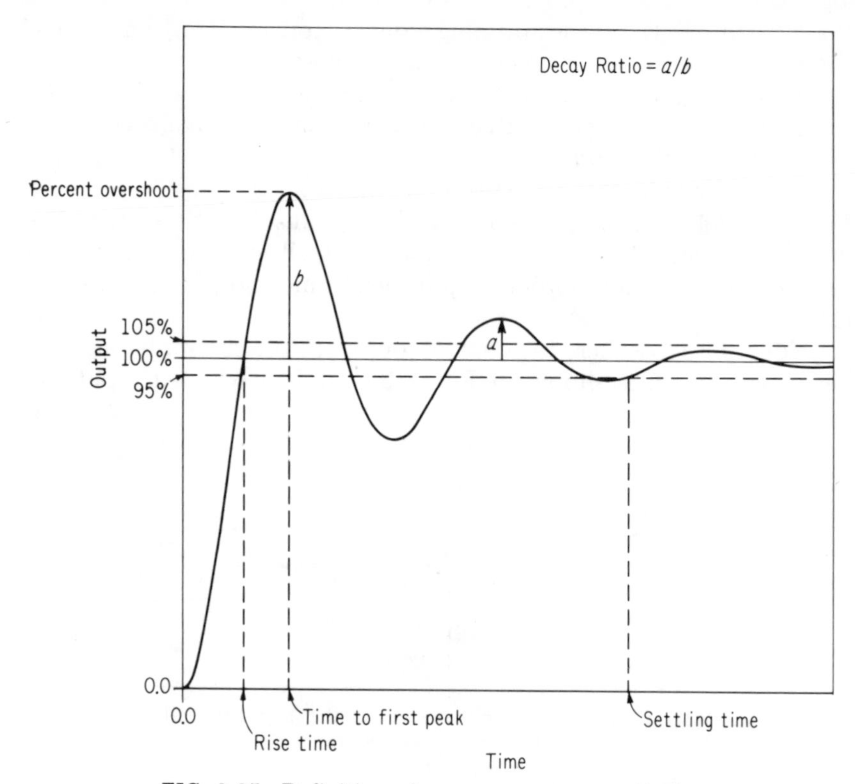

FIG. 6-25. Definition of common response criteria.

in the process industries for continuous controllers. Specifically, the one-quarter decay ratio has been recognized as a reasonable trade-off between a fast rise time and a reasonable settling time.

For discrete systems, this criterion can lead to some problems. Figure 6-26 shows a plot of the decay ratio vs. proportional gain for closed-loop control of a first-order-lag-plus-dead-time system (15). Note that in some cases ($\theta/\tau = 0.14$) there is more than one value of gain that gives a 1/4 decay ratio. Even so, because of its simplicity this criteria will continue to be a useful guide for tuning controllers.

The general form of the integral criteria is

$$I = \int_0^\infty f[t, e(t)]\, dt$$

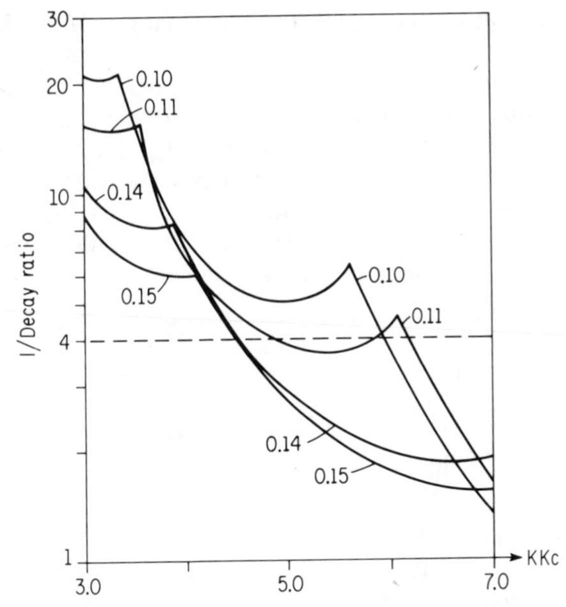

FIG. 6-26. Effect of loop gain on the decay ratio of the response of a first-order-lag-plus-dead-time system; $T/\tau = 0.4$; θ/τ is the parameter. (Reprinted by permission from Ref. 15.)

The most commonly used integral criteria are

1. Integral of the square error (ISE):

$$\text{ISE} = \int_0^\infty [e(t)]^2 \, dt$$

2. Integral of the absolute value of the error (IAE):

$$\text{IAE} = \int_0^\infty |e(t)| \, dt$$

3. Integral of time multiplied by the absolute value of the error:

$$\text{ITAE} = \int_0^\infty t\,|e(t)| \, dt$$

Other reasonable integral criteria can certainly be proposed.

For tuning purposes the objective is to select the values for the parameters in the control algorithm that minimize the value of the selected integral criterion for the response of the system to the selected input. Contrary to the shortcut criteria such as the decay ratio, they typically give a unique combination of settings for the optimum.

On the surface it might not appear possible to make general statements about the kind of responses that minimize each integral criterion. However, an examination of $f[t, e(t)]$ for each criterion indicates that they give different weights to certain errors. For example, large errors contribute more heavily to ISE than to IAE, meaning that ISE will favor responses with smaller overshoots for load changes, as illustrated in Fig. 6-27. Note that ISE gives a longer

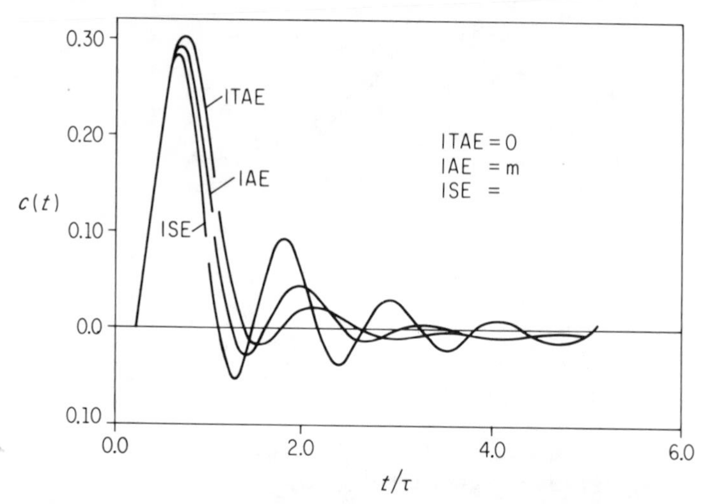

FIG. 6-27. Comparison of responses to a unit step change in load for different performance criteria; PI algorithm, $G(s) = (e^{-0.2\tau s})/(\tau s + 1)$; $T/\tau = 0.10$. (Reprinted by permission from Ref. 16.)

settling time. As time appears within the integral, ITAE weighs more heavily errors occurring late in time, and ignores to a large extent the unavoidable errors that occur early in time. As shown in Fig. 6-27, this criterion gives the shortest settling time but the most overshoot.

By far the most serious disadvantage of the integral criteria is the effort required to evaluate them. This can to some extent be offset in the digital control case because they can be readily calculated by the computer, giving possibilities of automated tuning techniques.

6-11 TUNING TECHNIQUES

The main difference between tuning continuous control loops and digital control loops is that in the latter case an additional parameter, the sampling time, must be taken into consideration.

Lopez (15,16) developed tuning graphs (based on integral criteria) for digital loops, but this additional parameter made it impossible to reduce them to convenient equation form. Lopez developed these graphs from repetitive solutions of the equations describing the control loop, searching for the tuning parameters that minimized the integral criteria.

Moore (17) examined the possibility of using the dead time equivalent to the sample and hold as developed in Sec. 5-3. The tuning techniques for continuous controllers relate the loop gain (KK_c), the ratio of the integral time to the time constant (T_i/τ), and the ratio of the derivative time to the time constant (T_d/τ) to the ratio of the process dead time to the time constant (θ/τ). If it is satisfactory to represent the sample-and-hold by a dead time equal to half the sampling time, the continuous tuning techniques could be applied by using $\theta + T/2$ instead of θ. Figure 6-28 compares the settings predicted in this manner to those calculated by Lopez for a PI controller. As they appear to be sufficiently accurate for tuning purposes, a review of continuous tuning techniques is in order.

The first tuning techniques to appear were all based on the one-quarter decay ratio criterion. The first of these was proposed by Ziegler and Nichols (1) in 1942. Cohen and Coon (18) presented a later version, followed finally by Smith and Murrill (19). The relationships for each of these are given in Table 6-3.

Lopez et al. (20) determined the control parameters that minimize the integral criteria ISE, IAE, and ITAE, the results appearing in Table 6-4. These as well as those presented by Smith and Murrill for the one-quarter decay ratio were calculated from step changes in the disturbance, which is what the typical analog controller is installed to compensate for. Rovira (21) extended the integral criteria tuning to set point changes, his results also appearing in Table 6-4.

Table 6-5 gives the results from applying the tuning techniques to the fourth-order process for which the first- and second-order models were derived in Sec. 6-1. The first-order model used was the Fit 3 model. This is the most conservative of the three, since the ratio θ/τ for this model was the largest of the three. Tuning parameters calculated using the other two fits gave tighter control and larger ITAE.

The values of the ITAE in Table 6-5 are the values obtained when the tuning parameters were applied to the original fourth-order process. Note that in each case considerable improvement can be obtained over the settings given by the first-order model, 107 percent (12.19 vs. 5.88) for set-point changes and 45 percent (13.84 vs. 9.59) for load changes. Therefore the results calculated from these methods are useful as only first approximations.

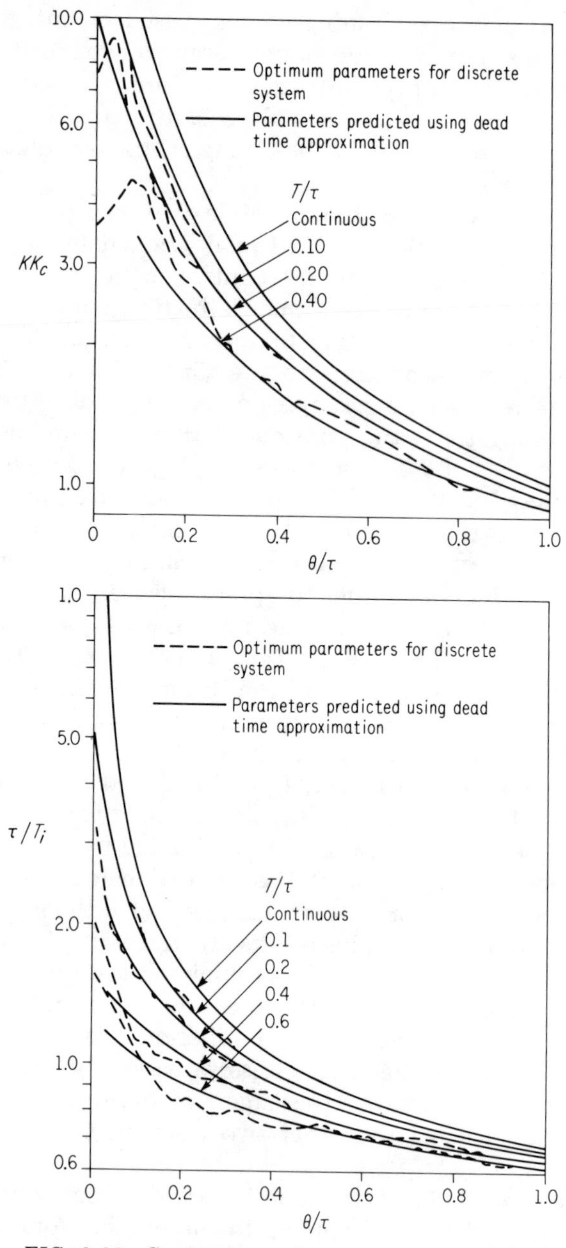

FIG. 6-28. Comparison of tuning parameters for a PI controller predicted by the dead time approximation to those of Lopez (16). (Reprinted by permission from Moore, Smith, and Murrill, "Simplifying Digital Control Dynamics for Controller Tuning and Hardware Lag Effects," *Instrument Practice*, January 1969.)

TABLE 6-3

Comparison of Ziegler-Nichols, Cohen-Coon, and $3C$ Equations (Reprinted by permission from Murrill, *Automatic Control of Processes*, Intext Educational Publishers, Scranton, Pa., 1967.)

Controller	Ziegler-Nichols	Cohen-Coon	$3C$
Proportional	$KK_c = (\theta/\tau)^{-1.0}$	$KK_c = (\theta/\tau)^{-1.0} + 0.333$	$KK_c = 1.208(\theta/\tau)^{-0.956}$
Proportional + Reset	$KK_c = 0.9(\theta/\tau)^{-1.0}$	$KK_c = 0.9(\theta/\tau)^{-1.0} + 0.082$	$KK_c = 0.928(\theta/\tau)^{-0.946}$
	$T_i/\tau = 3.33(\theta/\tau)$	$T_i/\tau = \dfrac{3.33(\theta/\tau)\,[1 + (\theta/\tau)/11.0]}{1.0 + 2.2(\theta/\tau)}$	$T_i/\tau = 0.928(\theta/\tau)^{0.583}$
Proportional + Reset + Rate	$KK_c = 1.2(\theta/\tau)^{-1.0}$	$KK_c = 1.35(\theta/\tau)^{-1.0} + 0.270$	$KK_c = 1.370(\theta/\tau)^{-0.950}$
	$T_i/\tau = 2.0(\theta/\tau)$	$T_i/\tau = \dfrac{2.5(\theta/\tau)\,[1.0 + (\theta/\tau)/5.0]}{1.0 + 0.6(\theta/\tau)}$	$T_i/\tau = 0.740(\theta/\tau)^{0.738}$
	$T_d/\tau = 0.5(\theta/\tau)$	$T_d/\tau = \dfrac{0.37(\theta/\tau)}{1.0 + 0.2(\theta/\tau)}$	$T_d/\tau = 0.365(\theta/\tau)^{0.950}$

TABLE 6-4

Tuning Relations Based on Integral Criteria (20, 21)

Form of tuning relation (unless otherwise noted)

$$Y = A \left(\frac{\theta}{\tau} \right)^B$$

where $Y = KK_c$ for proportional mode, τ/T_i for reset mode, T_d/τ for rate mode

A, B = constant for given controller and mode

θ, τ = pure delay time and first-order lag time constant based on the process reaction curve

Disturbance

Criterion	Controller	Mode	A	B
IAE	Proportional	Proportional	0.902	-0.985
ISE	Proportional	Proportional	1.411	-0.917
ITAE	Proportional	Proportional	0.490	-1.084
IAE	Proportional + Reset	Proportional	0.984	-0.986
		Reset	0.608	-0.707
ISE	Proportional + Reset	Proportional	1.305	-0.959
		Reset	0.492	-0.739
ITAE	Proportional + Reset	Proportional	0.859	-0.977
		Reset	0.674	-0.680
IAE	Proportional + Reset + Rate	Proportional	1.435	-0.921
		Reset	0.878	-0.749
		Rate	0.482	1.137
ISE	Proportional + Reset + Rate	Proportional	1.495	-0.945
		Reset	1.101	-0.771
		Rate	0.560	1.006
ITAE	Proportional + Reset + Rate	Proportional	1.357	-0.947
		Reset	0.842	-0.738
		Rate	0.381	0.995

Set Point

Criterion	Controller	Mode	A	B
IAE	Proportional + Reset	Proportional	0.758	-0.861
		Reset*	1.02	-0.323
ITAE	Proportional + Reset	Proportional	0.586	-0.916
		Reset*	1.03	-0.165
IAE	Proportional + Reset + Rate	Proportional	1.086	-0.869
		Reset*	0.740	-0.130
		Rate	0.348	0.914
ITAE	Proportional + Reset + Rate	Proportional	0.965	-0.855
		Reset*	0.796	-0.147
		Rate	0.308	0.929

*For this mode, equation should be of the form $Y = A + B(\theta/\tau)$

TABLE 6-5
Comparison of Controller Settings for Minimum ITAE

Method	K_c	$1/T_i$	T_d	ITAE, Set-Point Change	ITAE, Load Change
First-order tuning, load change	1.6001	0.3324	0.9228	20.97	13.84
Second-order tuning, load change	1.4501	0.4224	1.4328	16.48	10.73
True optimum, load change	1.2721	0.4104	1.4518	10.58	9.59
First-order tuning, set-point change	1.1198	0.2335	0.7546	12.19	24.65
Second-order tuning, set-point change	0.9101	0.3524	1.1628	5.93	15.73
True optimum, set-point change	0.9138	0.3465	1.1030	5.88	15.68

For the second-order model, the settings are much better, the possible improvement being less than one percent (5.93 vs. 5.88) for set-point changes and only 12 percent (10.73 vs. 9.59) for load changes. The settings in Table 6-5 for the second-order model were calculated using a search technique, although Lopez presented settings for load changes (14).

Another point that should be noted carefully is that settings that are optimal in some sense for load changes are not optimal for set-point changes, and vice versa. This is vividly illustrated in Fig. 6-29. Applying load settings to set-point changes leads to a more oscillatory behavior than desired; conversely, applying set-point settings to load changes leads to a more sluggish recovery than desired.

The culprit here is the integral mode. For set-point changes, the error becomes large the moment the set-point change is made. But for load changes, the error remains zero until the load change causes the process to begin to deviate from the set point. For this reason the value of $\int e\,dt$ tends to be larger for set-point changes, and thus the coefficient $1/T_i$ tends to be smaller. If the integral action is removed, the settings are consistent. Note that there is only one group of settings for a proportional controller in Table 6-4.

One solution to this dilemma is to store two sets of tuning parameters, one of which is used for a period of time after a set-point change is made and the other used at other times. A second solution is to make only small increments in the set point, i.e., ramp the set point to its new value.

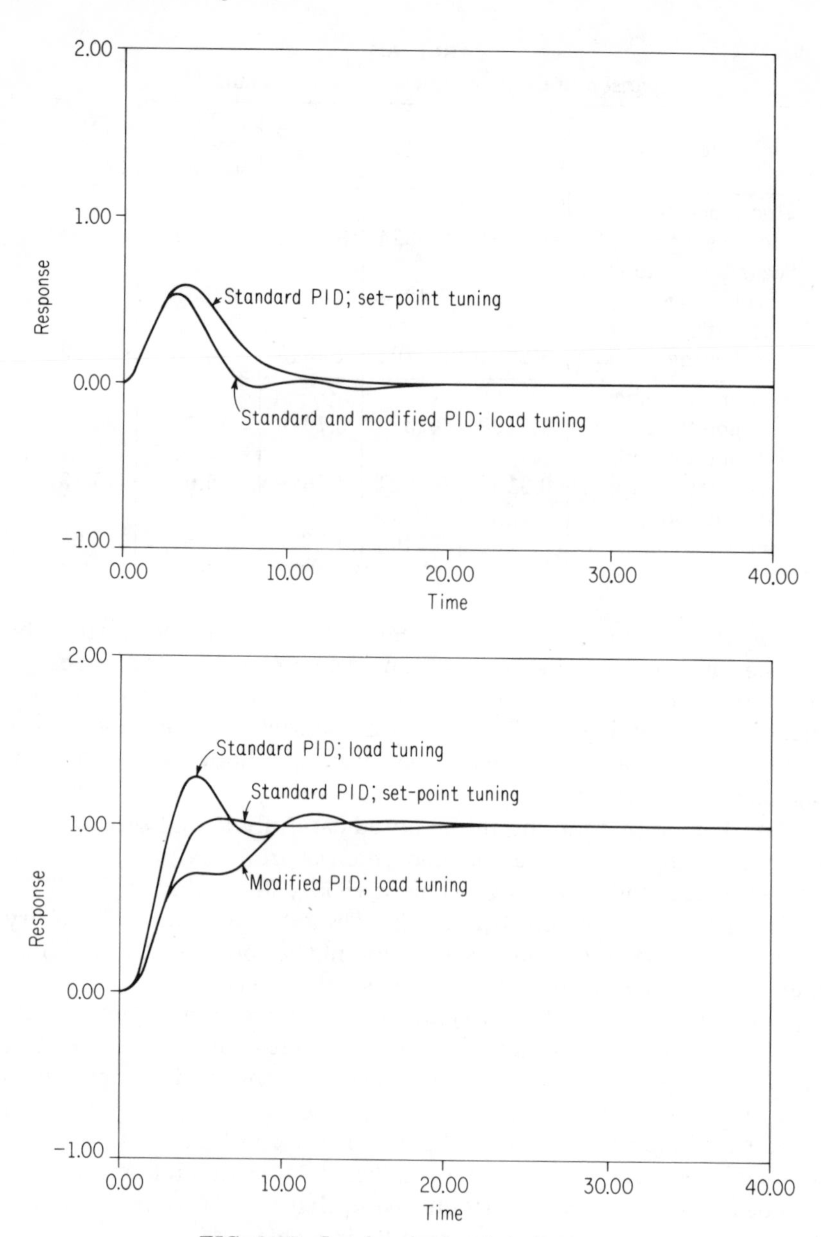

FIG. 6-29. Load and set-point responses.

A third solution is to modify the control algorithm. For the PI algorithm the velocity form is

$$\Delta m_n = K_c \left[(e_n - e_{n-1}) + \frac{T}{T_i} e_n \right]$$

Recalling that the error e_n is the set point r_n minus the controlled variable c_n, this equation is

$$\Delta m_n = K_c \left[(r_n - r_{n-1}) - (c_n - c_{n-1}) + \frac{T}{T_i} e_n \right]$$

The term $(r_n - r_{n-1})$ gives a proportional "kick" for set point changes.

But as illustrated in Fig. 6-29, a given set of tuning parameters appear to be tighter for set-point responses than for load responses. Thus it seems reasonable to eliminate this term and the corresponding proportional "kick," giving the algorithm

$$\Delta m_n = K_c \left[-(c_n - c_{n-1}) + \frac{T}{T_i} e_n \right]$$

This algorithm's behavior for load changes is identical to that of the original algorithm, since the term $(r_n - r_{n-1})$ is zero for load changes anyway.

The response of this algorithm is also shown in Fig. 6-29. The tuning parameters for load changes were used. Note that the set-point response is a little more sluggish that that of the original PI algorithm tuned to set point changes. This suggests the following approach (22):

1. If the controller's primary function is to compensate for load changes, the modified algorithm is attractive. This is the case in most continuous processes.
2. If the controller's primary function is to make set point changes, the standard algorithm should be used, but must be tuned for set-point changes. This is the case in many batch processes.
3. If both load and set-point changes are frequent, the modified algorithm tuned for load changes is probably the best choice.

The above discussion considers the PI case only for simplicity; the derivative mode can certainly be added.

6-12 SELECTION OF A SAMPLING TIME

The selection of the sampling time for a particular loop is generally not approached in a rigorous manner at present. For most algorithms, the control-loop performance improves (cost decreases) as the sampling time is decreased, as is illustrated in Fig. 6-30. However, there is a point of diminishing return, since the machine time and effort increase as the sampling time decreases. In fact, there should be an optimum sampling time for each loop, as illustrated in

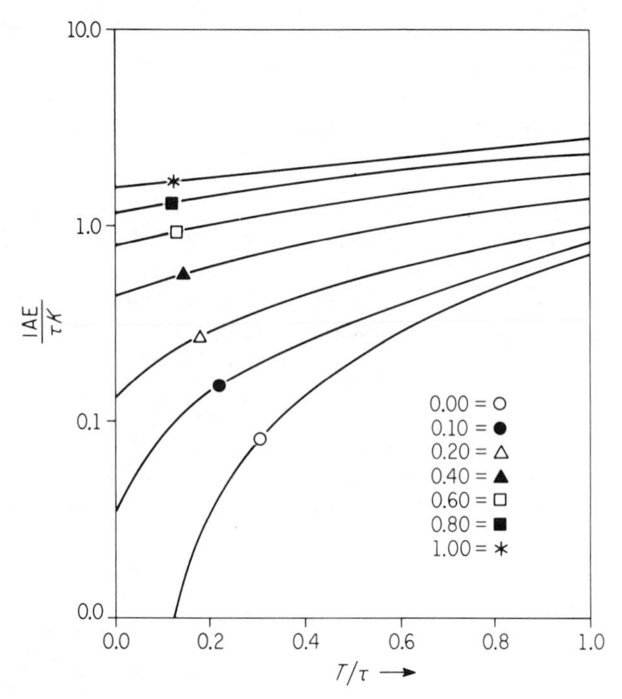

FIG. 6-30. Effect of sampling time on minimum error integral IAE criterion, PI control; θ/τ shown as parameter. (Reprinted by permission from Ref. 16.)

Fig. 6-31. Unfortunately, neither of these curves have been quantified very satisfactorily as of yet.

In early efforts in the DDC area, the following guide to the selection of sampling times was presented (23):

Type of Loop	Sampling Time, sec.
Flow	1
Level and pressure	5
Temperature	20

This reflects the general trend that flow loops are faster than level and pressure loops, and thus should be sampled faster. Temperature loops are generally the slowest loops of all, and therefore can be sampled slowest. However, there are exceptions to this, and it seems that the sampling time should rather be related to some process parameter.

For the conventional PI and PID control algorithms, the performance is inversely proportional to the parameter θ/τ, the ratio of

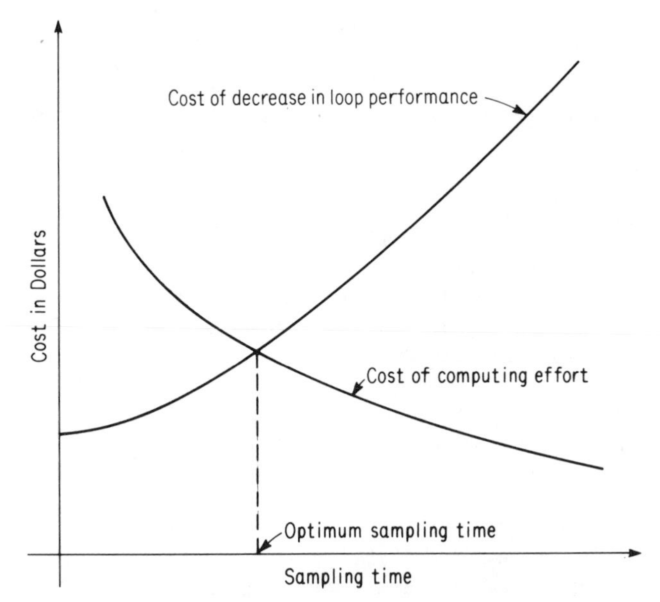

FIG. 6-31. Economic optimum sampling time.

process dead time to the process time constant. For sampled-data loops, this ratio should really be $(\theta + T/2)/\tau$, since the sample-and-hold behaves much like a dead time. Thus the loop performance in sampled-data systems is determined by the sum of two terms:

$$\frac{\theta}{\tau} + \frac{T/2}{\tau}$$

The first is due to the process; the second is due to the sampler-and-hold. When the magnitude of the term $(T/2)/\tau$ approaches the magnitude of θ/τ, the degradation of loop performance due to the sampler becomes significant. This type of reasoning supports the suggestion that the sampling time should be shorter than the process dead time (24).

In the case of algorithms derived by the use of z-transforms, the conclusion in the above paragraph tends to break down. Since the algorithm was derived based on a process model, the algorithm should compensate for any process dead time, assuming of course, that the model adequately describes the process. In this case the sampling time should be relatively independent of the dead time, but related in some manner to the process time constant.

In general, there is considerable room for improvement in the area of selection of the sampling time.

LITERATURE CITED

1. Ziegler, J. G., and N. B. Nichols, "Optimum Settings for Automatic Controllers," *Trans. ASME*, Vol. 64, No. 11 (November 1942), p. 759.
2. Miller, J. A., et al., "A Comparison of Controller Tuning Techniques," *Control Engineering*, Vol. 14, No. 12 (December 1967), p. 72.
3. Sten, J. W., "Evaluating Second-Order Parameters," *Instrumentation Technology*, Vol. 17, No. 9 (September 1970), pp. 39-41.
4. Oldenbourg, R. C., and H. Sartorius, *The Dynamics of Automatic Controls*, American Society of Mechanical Engineers, New York, 1948, p. 276.
5. Smith, O. J. M., "A Controller to Overcome Dead Time," *ISA J.*, Vol. 6, No. 2 (February 1959), pp. 28-33.
6. Cox, J. B., et al., "A Practical Spectrum of DDC Chemical-Process Control Algorithms," *ISA J.*, Vol. 13, No. 10 (October 1966), pp. 65-72.
7. Meyer, J. R., et al., "Simplifying Process Response Approximations," *Instruments and Control Systems*, Vol. 40, No. 12 (December 1967), pp. 76-79.
8. Kuo, B. C., *Analysis and Synthesis of Sampled-Data Control Systems*, Prentice-Hall, Englewood Cliffs, N.J., 1963.
9. Dahlin, E. B., "Designing and Tuning Digital Controllers," *Instruments and Control Systems*, Vol. 41, No. 6 (June 1968), p. 77.
10. Kalman, R. E., discussion following article entitled "Sampled-Data Processing Techniques for Feedback Control Systems," by A. R. Bergen and J. R. Ragazzini, *Trans. AIEE* (November 1954), pp. 236-247.
11. Mosler, H. A., L. B. Koppel, and D. R. Coughanowr, "Process Control by Digital Compensation," *A.I.Ch.E. J.*, Vol. 13, No. 4 (July 1967), p. 768.
12. Fertik, H. A., and C. W. Ross, "Direct Digital Control Algorithm with Anti-Windup Feature," *Leeds & Northrup Technical Journal* (January 1968), p. 1.
13. Gallier, P. W., and R. E. Otto, "Self-Tuning Computer Adapts DDC Algorithms," *Instrumentation Technology*, Vol. 15, No. 2 (February 1968), pp. 65-70.
14. Lopez, A. M., C. L. Smith, and P. W. Murrill, "An Advanced Tuning Method," *British Chemical Engineering*, Vol. 14, No. 11 (November 1969), pp. 1552-1555.
15. Lopez, A. M., P. W. Murrill, and C. L. Smith, "Optimal Tuning of Proportional Digital Controllers," *Instruments and Control Systems*, Vol. 41, No. 10 (October 1968), p. 97.
16. ________, "Tuning PI and PID Digital Controllers," *Instruments and Control Systems*, Vol. 42, No. 2 (February 1969), p. 89.
17. Moore, C. F., C. L. Smith, and P. W. Murrill, "Simplifying Digital Control Dynamics for Controller Tuning and Hardware Lag Effects," *Instrument Practice*, Vol. 23, No. 1 (January 1969), p. 45.
18. Cohen, G. H., and G. A. Coon, "Theoretical Considerations of Retarded Control," Taylor Instrument Companies Bulletin TDS-10A102.
19. Smith, C. L., and P. W. Murrill, "Controllers—Set Them Right," *Hydrocarbon Processing and Petroleum Refiner*, Vol. 45, No. 2 (February 1966), pp. 105-124.

20. Lopez, A. M. et al., "Controller Tuning Relationships Based on Integral Performance Criteria," *Instrumentation Technology*, Vol. 14, No. 11 (November 1967), p. 57.
21. Rovira, A. A., P. W. Murrill, and C. L. Smith, "Tuning Controllers for Setpoint Changes," *Instruments and Control Systems*, Vol. 42, No. 12 (December 1969), p. 67.
22. Rovira, A. A., "Modified PI Algorithm for Digital Control," *Instruments and Control Systems*, Vol. 43, No. 8 (August 1970), pp. 101–102.
23. ____, Minutes, Users Workshops on Direct Digital Control, April 3–4, 1963, Princeton, N.J.; May 6–7, 1964, Princeton, N.J.; Oct. 1–2, 1965, Anaheim, Calif.
24. Skinskey, F. G., *Process Control Systems*, McGraw-Hill, New York, 1967.

chapter 7

On-Line Identification Techniques

Experimental modeling techniques generally fall into two categories: frequency-domain techniques and time-domain methods. Each has its advantages and disadvantages. For example, virtually the first step in most time-domain methods is to postulate a process model. Parameters in the model are relatively easy to determine, but little guidance is available for improving the form of the model. In the frequency-domain methods, the model does not appear until virtually the last step, and considerable direction is generally available as to the proper form of the model.

However, frequency-domain techniques cannot be as readily applied on-line as can time-domain methods. Since on-line methods are the primary subject of this chapter, we shall consider only time-domain methods. Reference 1 gives a thorough discussion of frequency domain methods.

Throughout this chapter we shall emphasize the utility of a discrete model in formulating a fast, efficient identification method that can be used on-line. Generally these methods will be tailored to a specific model in order to achieve this speed and efficiency. Thus the generality so important in off-line identification packages is frequently lost.

7-1 GENERAL APPROACH TO TIME DOMAIN IDENTIFICATION

Identification in the time domain generally reduces to a nonlinear regression problem, although in special cases linear regression may apply. The method proceeds basically as follows.

1. Make a process "test," recording the observed process input(s) and output(s). Conceptually, this "test" could be normal operating data, but best results are obtained using a more substantial input whenever practical. In the discussion in this chapter we will restrict our attention to processes with one input and one output.
2. Postulate a model for the process. It is certainly desirable to have available from other sources some indication of the probable order of the process, of the existence of an appreciable dead time (and hopefully, an estimate of its magnitude), the general form of any appreciable nonlinearity, and so on. Refinements can certainly be made upon this information, but the better this initial information the easier the identification.
3. With assumed values for the unknown parameters, solve the model equations and compare the resulting solution to the experimentally observed response. The assumed values can be updated and the solution repeated until the model response matches the experimental response in some "best" sense.

This procedure is illustrated in Fig. 7-1.

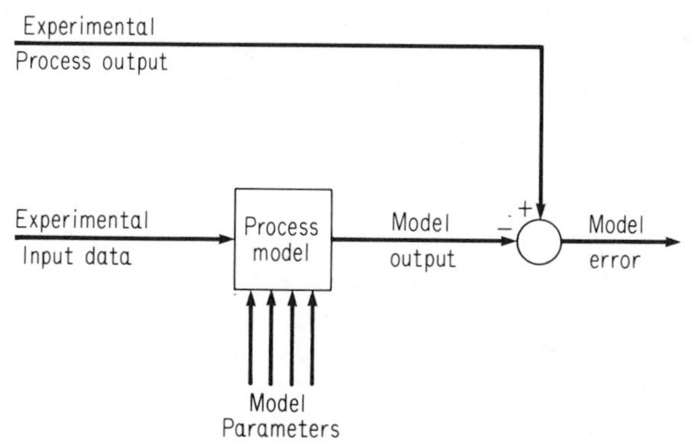

FIG. 7-1. Illustration of time-domain identification procedure.

Since most processes are continuous, the true process model generally contains one or more differential equations. But when designing sampled-data control systems by z-transforms, the first step is to convert the continuous process transfer function into a pulse transfer function or discrete model. This step could be eliminated completely if a discrete model rather than a continuous model were posed initially. Other advantages of using the discrete model are (1) that fewer calculations are required to compute the process response from the experimental input data, and (2) that at least in certain cases linear regression can be used instead of nonlinear regression to compute the model parameters that give the "best" fit.

Now consider the criterion for ascertaining which model response gives the "best" fit to the experimental response. If we let $e(t)$ be the difference between the model response and the experimental response at time t, or e_i be the difference at the ith observation, the following possibilities exist:

1. *Sum of squares*, $\int [e(t)]^2 \, dt$ or $\Sigma \, e_i^2$

This criterion penalizes for large deviations heavier than for small ones. This criterion is undoubtedly the most commonly used, probably as a carryover from linear regression.

2. *Sum of absolute values*, $\int [|e(t)|] \, dt$ or $\Sigma \, |e_i|$

This criterion penalizes equally for both large and small deviations. Ritter (2) compared this criterion to the sum of squares criterion on a relatively simple example and found little difference in behavior.

3. *Minimize the maximum deviation*, $\min \{\max[e(t)]\}$ or $\min \{\max[e_i]\}$

While a small region of poor fit may be overshadowed by large regions of good fit when using a sum, this criterion is determined solely by the point at which the model deviates most from the experimental response.

The selection of the criterion function certainly affects the final values of the parameters in the model, although the values will probably not be very different if the model is reasonably good.

One perplexing aspect of this approach is that the values of the parameters will be different for different process inputs (even if the process is linear), unless the model is exact. Frequency response on the other hand does not suffer this disadvantage if the process is linear.

The design of control systems is unquestionably easier if the model is linear. But since most processes are nonlinear to some degree, they cannot be represented over wide ranges by linear models. This means that for a linear model to be adequate, it must be determined at or near the usual operating level of the nonlinear process. If the operating level changes appreciably, it will be necessary to reevaluate the parameters in the model. This makes on-line identification techniques very desirable.

7-2 ESTIMATING PARAMETERS BY LINEAR REGRESSION

Unless a system contains some known nonlinearity, the first attempt is to represent it by a linear transfer function of the type

$$G(s) = \frac{\alpha_m s^m + \cdots + \alpha_1 s + \alpha_0}{a_n s^n + \cdots + a_1 s + 1} \tag{7-1}$$

In many cases dead time is also included, as described in a later section. If $u(t)$ is the input to the system and $x(t)$ is the output, Eq. 7-1 can alternatively be represented by the following differential equation:

$$\begin{aligned} a_n \frac{d^n x(t)}{dt^n} + \cdots + a_1 \frac{dx(t)}{dt} + x(t) \\ = \alpha_m \frac{d^m u(t)}{dt^m} + \cdots + \alpha_1 \frac{du(t)}{dt} + \alpha_0 u(t) \end{aligned} \tag{7-2}$$

In either case, the identification problem reduces to determining values for coefficients $a_n, \ldots, a_1, \alpha_m, \ldots, \alpha_1, \alpha_0$.

When using Eq. 7-2 directly as the model, a search technique or some other nonlinear regression approach is required to determine the parameter values that minimize one of the fit criteria given in the previous section. But by transforming Eq. 7-2 into a difference equation or discrete model, linear regression can often be used. Computationally, this is definitely advantageous.

Let x_i and u_i be the data obtained by sampling the continuous functions $x(t)$ and $u(t)$ every T units of time. As x_i and u_i are number (or pulse) sequences, they should really be related by a pulse transfer function rather than by the continuous relationship of either Eq. 7-1 or 7-2. Recall that the pulse transfer function relates the pulse sequences at two samplers in a sampled data system. Figure 7-2 shows the block diagram for deriving the pulse transfer function to be used in the identification procedure.

To show that at least in some cases the identification can be accomplished with linear regression, consider the specific case in which the system is assumed to be first-order. Assuming the hold to be a zero-order hold, the system is as illustrated in Fig. 7-1b. The pulse transfer function is therefore given by

$$HG(z) = \mathfrak{z}\left[\frac{1 - e^{-sT}}{s} \cdot \frac{K}{\tau s + 1}\right] = \frac{K(1 - e^{-T/\tau}) z^{-1}}{1 - z^{-1} e^{-T/\tau}} = \frac{X(z)}{U(z)} \tag{7-3}$$

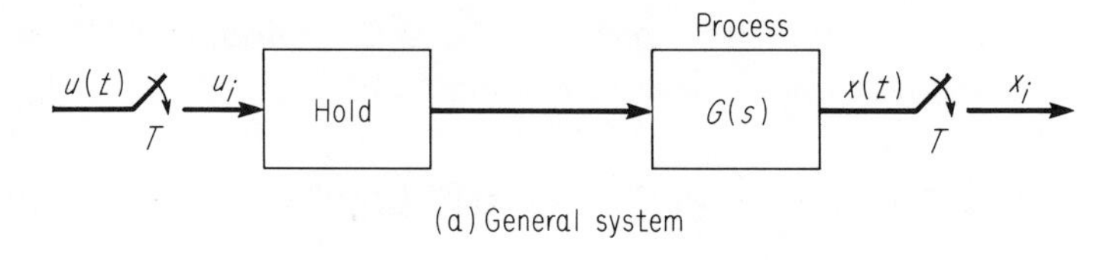

(a) General system

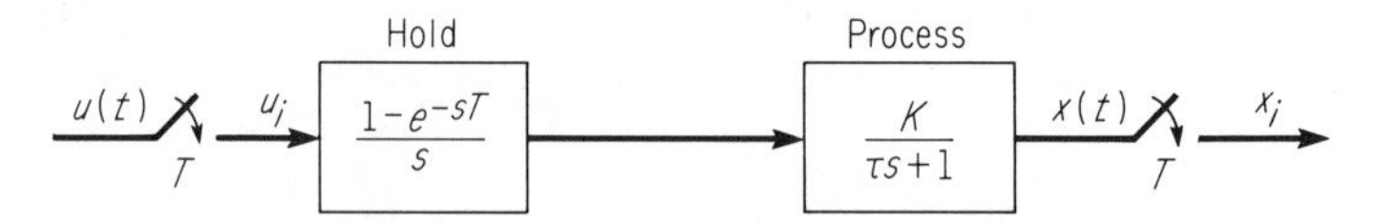

(b) First-order system with zero-order hold

FIG. 7-2. Block diagram for deriving pulse transfer function.

In terms of sequences x_i and u_i, this relationship is

$$x_{i+1} = e^{-T/\tau} x_i + K(1 - e^{-T/\tau}) u_i \tag{7-4}$$

This equation does not quite tell the whole story. At steady-state, $u_i = u_s$, $x_i = x_{i+1} = x_s$, and the equation reduces to

$$x_s = Ku_s \tag{7-5}$$

where x_s is the equilibrium output corresponding to u_s. Generally, it is necessary to add a constant bias, say D, to Eq. 7-4, giving

$$x_{i+1} = e^{-T/\tau} x_i + K(1 - e^{-T/\tau}) u_i + D \tag{7-6}$$

Now the steady-state relationship becomes

$$x_s = Ku_s + D'$$

where $D' = D/(1 - e^{-T/\tau})$.

Another way to visualize D' is as the steady-state value of x corresponding to a zero input.

For convenience, we will introduce the parameters a and b so that Eq. 7-6 becomes

$$x_{i+1} = ax_i + bu_i + D \tag{7-6a}$$

where
$$a = e^{-T/\tau}$$
$$b = K(1 - e^{-T/\tau})$$
$$D = D'(1 - e^{-T/\tau})$$

In formulating the identification procedure, two avenues are open. First x_{i+1} could be replaced by the model output, namely $\hat{x}_{i+1}$, to give

$$\hat{x}_{i+1} = ax_i + bu_i + D \tag{7-7}$$

The model output at the next sampling instant is calculated from the current process output x_i and current input u_i. A model used in this fashion is often referred to as a one-step-ahead predictor.

Alternatively, both x_{i+1} and x_i could be replaced by the model output, namely $\hat{x}_{i+1}$ and $\hat{x}_i$, respectively, to give

$$\hat{x}_{i+1} = a\hat{x}_i + bu_i + D \qquad (7\text{-}8)$$

That is, the model output at the next sampling instant is calculated from the current input and the model output for the current sampling instant. The calculations must be initialized with a value for $\hat{x}_0$. Note that the model output is in no way dependent upon the true process output, and such a model is sometimes called "free-running."

Although similar, the models in Eqs. 7-7 and 7-8 are not equivalent. For the same set of data for x_i and u_i, minimization of the same fit criterion will give different values for the parameters a, b, and D. To determine the parameters in Eq. 7-8, nonlinear regression must be used. However, linear regression can be used for Eq. 7-7.

Using the residual sum of squares Σe_i^2 as the fit criterion for Eq. 7-7, its value is given by

$$\begin{aligned}\Sigma e_i^2 &= \Sigma(x_{i+1} - \hat{x}_{i+1})^2 \\ &= \Sigma(x_{i+1} - ax_i - bu_i - D)^2\end{aligned}$$

The best values of a, b, and D are the values that minimize this expression. At the minimum, the partials with respect to each of the parameters must equal zero:

$$\frac{\partial}{\partial a}[\Sigma e_i^2] = -2\Sigma x_i(x_{i+1} - ax_i - bu_i - D) = 0$$

$$\frac{\partial}{\partial b}[\Sigma e_i^2] = -2\Sigma u_i(x_{i+1} - ax_i - bu_i - D) = 0$$

$$\frac{\partial}{\partial D}[\Sigma e_i^2] = -2\Sigma(x_{i+1} - ax_i - bu_i - D) = 0$$

This gives three equations and three unknowns (N = total number of data points):

$$\begin{aligned}a\Sigma x_i^2 + b\Sigma x_i u_i + D\Sigma x_i &= \Sigma x_{i+1} x_i \\ a\Sigma x_i u_i + b\Sigma u_i^2 + D\Sigma u_i &= \Sigma x_{i+1} u_i \\ a\Sigma x_i + b\Sigma u_i + ND &= \Sigma x_{i+1}\end{aligned} \qquad (7\text{-}9)$$

or

$$a\Sigma x_i/N + b\Sigma x_i u_i/N + D\Sigma x_i/N = \Sigma x_{i+1} x_i/N$$

$$a\Sigma x_i u_i/N + b\Sigma u_i^2/N + D\Sigma u_i/N = \Sigma x_{i+1} u_i/N \tag{7-9a}$$
$$a\Sigma x_i/N + b\Sigma u_i/N + D = \Sigma x_{i+1}/N$$

These equations can be solved by any number of techniques. Alternatively, a linear regression package can be used since these equations are identical to those encountered in the usual linear regression formulation.

An alternative approach to arriving at an expression analogous to Eq. 7-6 is to use finite differences instead of the continuous derivatives in the differential equation equivalent to the first-order lag transfer function:

$$\tau \frac{dx(t)}{dt} + x(t) = Ku(t) + D'$$

where again D' is the bias. Inserting finite differences yields

$$\tau \frac{(x_{i+1} - x_i)}{T} + x_i = Ku_i + D'$$

or

$$x_{i+1} = \left(1 - \frac{T}{\tau}\right)x_i + \frac{KT}{\tau} u_i + \frac{D'T}{\tau} \tag{7-10}$$

This equation is also linear, being quite similar to Eq. 7-6. Equation 7-7 can again be used for the model, the parameters being related to the coefficients as follows:

$$\begin{aligned} D &= D'T/\tau \\ a &= 1 - T/\tau \\ b &= KT/\tau \end{aligned} \tag{7-11}$$

As contrasted to the z-transform approach, this method is only strictly applicable when the sampling time T is small enough for the finite differences to be reasonably accurate. But as we shall see in the next section, the z-transform approach will only yield a linear regression formulation when the number of poles exceeds the number of zeroes by exactly one. The finite difference approach does not suffer this disadvantage.

The constant term D can be eliminated by formulating the algorithm in terms of differences instead of actual values or deviations. Writing Eq. 7-7 for $\hat{x}_i$ in terms of x_{i-1} and u_{i-1}, and then subtracting from Eq. 7-7 gives

$$(\hat{x}_{i+1} - \hat{x}_i) = a(x_i - x_{i-1}) + b(u_i - u_{i-1}) \tag{7-12}$$

The linear regression equations 7-10 with D equal zero apply, but the equation derived from $\frac{\partial}{\partial D}\,[\Sigma e_i^2]$ should be deleted.

Alternatively, Eq. 7-7 can be summed starting with the first point and continuing through the equation for x_{i+1} to give

$$\sum_{j=1}^{i+1} x_j = a\sum_{j=0}^{i} x_j + b\sum_{j=0}^{i} u_j + iD \qquad (7\text{-}13)$$

Dividing through by i gives

$$\overline{X}_{i+1} = a\overline{X}_i + b\overline{U}_i + D \qquad (7\text{-}14)$$

where

$$\overline{X}_{i+1} = \frac{1}{i}\sum_{j=1}^{i+1} x_j$$

$$\overline{X}_i = \frac{1}{i}\sum_{j=0}^{i} x_j$$

$$\overline{U}_i = \frac{1}{i}\sum_{j=0}^{i} u_j$$

This method is essentially based upon the integral of the observed data.

If the system also contains dead time, the first-order transfer function would be

$$G(s) = \frac{Ke^{-\theta s}}{\tau s + 1}$$

If θ is an integer multiple of the sampling time, say nT, then Eq. 7-11 becomes

$$x_{i+1} = ax_i + bu_{i-n} + D \qquad (7\text{-}15)$$

As before, the parameters a, b, and D can be related to the continuous parameters K, τ, and D', depending upon whether z-transforms or differences are used to derive the discrete model.

Since the dead time appears in the difference equation as part of the subscript in the term u_{i-n}, linear regression cannot be used to determine a value for this parameter. However, if a value for n is assumed, linear regression i.e., Eq. 7-9, can still be used to determine values for a, b, and D. The procedure is as follows:

1. Assume a value for n.
2. Use linear regression to obtain values for a, b, and D, and the residual sum of squares.
3. Repeat the previous steps for different values of n until the value of n is found that gives the least residual sum of squares. This value of n and the corresponding values of a, b, and D are the best values.

This last step in essence requires a one-dimensional search.

Dahlin (3) studied a system, specifically a Fourdrinier paper machine, for which the experimental data contained a significant amount of noise. He did not consider Eq. 7-12 (based on differences), apparently because the noise would have contributed heavily to these differences. For this case, the method based on Eq. 7-14 was very attractive because of the excellent filtering capabilities of the integrals. In order to determine the bias, Dahlin took a record of data while holding the input constant instead of letting the bias be an unknown in the regression formulation.

The characteristics of the two methods as found in this study are given below.

Errors in estimation of a and b:
  Eq. 7-8: Quite high in the presence of disturbances, but independent of test pulses.
  Eq. 7-14: Small if tests are not too far apart and if the pole is small.
Errors in estimation of dead time:
  Eq. 7-8: Very small, particularly if test pulses have fast rise time and the time constant is small.
  Eq. 7-14: Large in presence of disturbances.
Effect of errors in estimate of bias D:
  Eq. 7-8: Parameter errors are small and independent of length of experiment.
  Eq. 7-14: Parameter errors increase rapidly with length of experiment.
Divergence of solution:
  Possible with both methods for a given set of data, but both do not necessarily diverge for the same set of data.
Effect of prefiltering the data:
  Eq. 7-8: Excellent ability to improve parameter estimates
  Eq. 7-14: Prefiltering is of value.

Dahlin suggested a "hybrid" method which relied upon both methods. He indicates that this greatly enhances the possibilities for convergence as the strengths of one method offset the weaknesses of the other. As for filtering, he suggests an exponential filter whose time constant is $2.49a$.

7-3 DEVELOPMENT OF A NONLINEAR "LEAST SQUARES" REGRESSION (4)

Consider the development of a nonlinear least-squares program to fit a set of input-output data to a second-order lag plus dead-time model of the form.

$$G(s) = \frac{Ke^{-\theta s}}{(\tau_1 s + 1)(\tau_2 s + 1)} = \frac{X(s)}{U(s)} \tag{7-16}$$

where K = gain
τ_1, τ_2 = time constants (real, distinct)
θ = dead time (real)

As before, the first step is to express $\hat{x}_i$ in terms of the pulse transfer function corresponding to the dynamic model (again a zero-order hold is assumed):

$$HG(z) = \mathfrak{z}\left[\frac{1 - e^{-sT}}{s} \cdot \frac{Ke^{-\theta s}}{(\tau_1 s + 1)(\tau_2 s + 1)}\right] = \frac{Kz^{-n}(b_1 z^{-1} + b_2 z^{-2})}{1 - a_1 z^{-1} + a_2 z^{-2}} \tag{7-17}$$

where $n = \theta / \tau$, assuming θ is an integer multiple of T,

$$b_1 = (\tau_2 - \tau_1 + \tau_1 e^{-T/\tau_1} - \tau_2 e^{-T/\tau_2})/(\tau_2 - \tau_1) \tag{7-18}$$

$$b_2 = [(\tau_2 - \tau_1)e^{-T/\tau_1} e^{-T/\tau_2} + \tau_1 e^{-T/\tau_2} - \tau_2 e^{-T/\tau_1}]/(\tau_2 - \tau_1) \tag{7-19}$$

$$a_1 = e^{-T/\tau_1} + e^{-T/\tau_2} \tag{7-20}$$

$$a_2 = e^{-T/\tau_1} e^{-T/\tau_2} \tag{7-21}$$

Expressing as a difference equation gives the following discrete model:

$$x_i = a_1 x_{i-1} - a_2 x_{i-2} + b_i u_{i-(n+1)} + b_2 u_{i-(n+2)} \tag{7-22}$$

For sake of simplicity we shall assume that the x's and u's are measured from some known equilibrium position. If not, an unknown bias can be included. Furthermore, if it is not satisfactory to restrict θ to be an integer multiple of the sampling time, the modified z-transform can be used. This gives another input term in Eq. 7-22 and an additional parameter.

The objective of this regression is to find values of τ_1, τ_2, θ, and K which give the minimum mean least-squares fit-error E, subject to the constraints that τ_1, τ_2, and θ are positive and real:

$$E = \frac{1}{N}\sum_{i=1}^{N} (x_i - a_1 x_{i-1} + a_2 x_{i-2} - b_1 u_{i-(n+1)})^2 \tag{7-23}$$

If the value of n were known, the coefficients a_1, a_2, b_1, and b_2 could be determined via linear regression. This offers the possibilities of performing a one-dimensional search on n, evaluating the remaining coefficients at each step. This is essentially the approach outlined in the last section.

But suppose values of τ_1, τ_2, K, and θ must be determined (i.e., the discrete model in Eq. 7-17 is not satisfactory). A natural suggestion is to determine values for a_1, a_2, b_1, and b_2 with the approach outlined in the previous paragraph, and then solve Eqs. 7-18 through 7-21 for τ_1, τ_2, and K. However, we have four equations in three unknowns, an overdetermined set. The alternative is to perform the minimization of the fit error E defined in Eq. 7-23 by performing a direct search over τ_1, τ_2, K, and θ using a multidimensional search technique such as Pattern(5).

Regardless of which parameters are to be identified (i.e., a_1, a_2, b_1, b_2, τ_1, τ_2, K and θ), some preliminary manipulations can lead to a formulation requiring considerably less core storage and less computational time. These two considerations, of course, are of paramount importance in on-line applications. To begin these manipulations, Eq. 7-23 can be expanded by squaring the terms in brackets to yield

$$E = SXX - 2SXU + SUU \tag{7-24}$$

where the terms

$$SXX = (\Sigma x_i^2 - 2a_1 \, \Sigma x_{i-1} x_i + 2a_2 \, \Sigma x_{i-2} x_i + a_1^2 \, \Sigma x_{i-1}^2 - 2a_2 a_1 \, \Sigma x_{i-2} x_{i-1} + a_2^2 \, \Sigma x_{i-2}^2)/N \tag{7-25}$$

$$SXU = (b_1 \, \Sigma u_{i-n-1} x_i - b_2 \, \Sigma u_{i-n-2} x_i - a_1 b_1 \, \Sigma u_{i-n-1} x_{i-1} - a_1 b_2 \, \Sigma u_{i-n-2} x_{i-1} - a_2 b_1 \, \Sigma u_{i-n-1} x_{i-2} - a_2 b_2 \, \Sigma u_{i-n-2} x_{i-2})/N \tag{7-26}$$

$$SUU = (b_1^2 \, \Sigma u_{i-n-1}^2 + 2b_2 b_1 \, \Sigma u_{i-n-2} u_{i-n-1} + b_2^2 \, \Sigma u_{i-n-2}^2)/N \tag{7-27}$$

The minimization of the mean residual sum of squares must be undertaken by a search technique of some type. All such search techniques require numerous evaluations of the residual sum of squares for various values of τ_1, τ_2, K, and θ. The approach will be to calculate the values of a_1, a_2, b_1, and b_2 from τ_1, τ_2, and K via Eqs. 7-18 through 7-21. The mean residual sum of squares can be calculated by Eq. 7-24 with the aid of Eqs. 7-25, 7-26, and 7-27. The number of computations involved must be kept to a minimum if it is to be usable on-line.

Judging from the number of summation terms in Eqs. 7-25,

7-26, and 7-27, it might appear that instead of shortening the time required for each evaluation, it has instead been greatly increased. However, it will be shown that each summation term can be reduced to a constant which remains the same throughout the search, thereby greatly reducing the computations required for each iteration.

Immediately it is possible to reduce the summation terms contained in Eqs. 7-25 and 7-27 to constants. They are in no way dependent on any of the search parameters (τ_1, τ_2, θ, and K), and need only be evaluated at the beginning of the search. The Σx_i^2, $\Sigma x_i x_{i-1}$, $\Sigma x_i x_{i-2}$, Σx_{i-1}^2, etc., terms can easily be accumulated in the data collection program during the actual test period. Similarly, the summation terms involving only products of u can be determined noting that the following relationships should hold for large sets of data:

$$\Sigma u_{i-n-1}^2 = \Sigma u_{i-n-2}^2 = \Sigma u_{i-n-3}^2, \quad \text{etc.}$$

$$\Sigma u_{i-n-1} u_{i-n-2} = \Sigma u_{i-n-2} u_{i-n-3}, \quad \text{etc.}$$

The summation terms contained in Eq. 7-26 cannot be reduced to constants quite as easily as the above terms. Note that all the terms in Eq. 7-26 contain cross-products of x and u ($\Sigma u_{i-n-1} x_i$, $\Sigma u_{i-n-2} x_i$, etc.). Unlike the summation terms containing only x or u, the numerical value of summations involving cross-products depends on the value of n (determined by the dead time) which varies from iteration to iteration in the multidimensional search. However, at least an estimate of high and low bounds on the dead time are generally known a priori. It is thus possible to accumulate the cross-product terms in Eq. 7-26 into a vector $\overline{S}$ defined as follows:

$$\begin{aligned}
&\Sigma u_{i-n-1} x_{i-2} & & & &= S_{n+1} \\
&\Sigma u_{i-n-1} x_{i-1} &= \Sigma u_{i-n-2} x_{i-2} & & &= S_n \\
&\Sigma u_{i-n-1} x_i &= \Sigma u_{i-n-2} x_{i-1} &= \Sigma u_{i-n-3} x_{i-2} & &= S_{n-1} \\
& & \Sigma u_{i-n-2} x_i &= \Sigma u_{i-n-3} x_{i-1} & &= S_{n-2} \\
& & & \Sigma u_{i-n-3} x_i & &= S_{n-3}
\end{aligned}$$

At this point the development of a fast, efficient regression program is essentially complete. All the summation terms indicated in Eqs. 7-25, 7-26, and 7-27 have been reduced to constants thereby greatly reducing the time required for each iteration in the search. The need for lengthy data tables has also been eliminated, requiring instead a vector $\overline{S}$ which is considerably shorter in length. In general, the form of the regression appears to be more in line with the requirements demanded by on-line applications.

7-4 EXPONENTIALLY MAPPED ESTIMATES

In the previous sections, several summation terms appeared in the regression formulations. This section is devoted to presenting an alternative to the direct evaluation of the summation terms that is potentially attractive on process control computers (6).

Suppose we begin by examining the term $\Sigma x_i^2 / N$ in Eq. 7-9 or 7-25. An alternative to a direct evaluation of this term is to use a weighted estimate of the average instead of the true arithmetic average. The general definition of the continuous weighted average of the function over the interval t_1 to t_2 is given by

$$\overline{x}^2(t) = \frac{\int_{t_1}^{t_2} x(t)^2 \varphi(t)\, dt}{\int_{t_1}^{t_2} \varphi(t)\, dt} \tag{7-28}$$

Suppose we chose the weighting function $\varphi(t)$ to be the exponential $e^{-\alpha t}$, the lower limit t_1 to be $-\infty$, and the upper limit to be t. Thus the function $\overline{x}^2(t)$ is given by

$$\overline{x}^2(t) = \alpha e^{-\alpha t} \int_{-\infty}^{t} x(\tau)^2 e^{\alpha \tau}\, d\tau$$

$$= \alpha \int_{-\infty}^{t} x(\tau)^2 e^{-\alpha(t-\tau)}\, d\tau \tag{7-29}$$

This equation is nothing more than the convolution of the function $x(t)^2$ with the exponential $e^{-\alpha t}$. If $F(s)$ is the Laplace transform of $x(t)^2$ and $\overline{F}(s)$ of $\overline{x}^2(t)$, Eq. 7-29 would be expressed by Laplace transforms as

$$\overline{F}(s) = F(s)\left(\frac{\alpha}{s+\alpha}\right) \tag{7-30}$$

or in block diagram form as

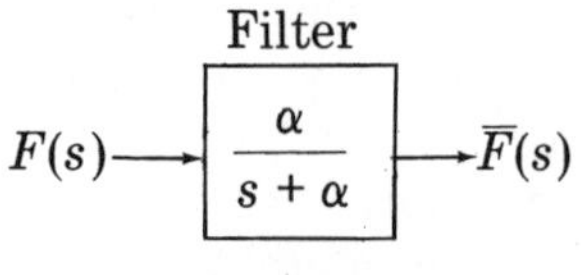

Filter

$$F(s) \longrightarrow \boxed{\frac{\alpha}{s+\alpha}} \longrightarrow \overline{F}(s)$$

Consequently $\overline{x}^2(t)$ can be obtained by passing $x(t)^2$ through a first-order lag filter.

As for programming, this can be readily accomplished in Fortran with only one statement. Recall that $x(t)^2$ would be available to the computer as a sampled sequence x_i^2. The sequence $\overline{x}_i^2$ could be obtained from x_i^2 by the recurrence relationship

$$\overline{x}_i^2 = Qx_i^2 + (1 - Q)\overline{x}_i^2 \tag{7-31}$$

where the parameter Q is related to the filter pole α by

$$Q = 1 - e^{-\alpha T} \tag{7-32}$$

Only the current values of x_i^2 and $\overline{x}_i^2$ need to be stored.

If this approach is used for all the summation terms in equations such as Eq. 7-9, several advantages acrue.

1. In the direct evaluation of the sums, very large values can occur. If fixed (integer) arithmetic is used to conserve computation time, the risk of overflow is quite serious. With the weighted average, overflow is no problem.
2. Using the direct-summation approach, the previous average must be discarded at the beginning of a new set of data. The new average is not available until all the data has been taken. Using the weighted average, the estimate is available at all times, providing some continuity between data sets. This would also allow the regression to be performed as frequently as desired.
3. Figure 7-3 presents a pictorial representation of the weighting function. The most recent information is weighted heaviest, which seems reasonable in an on-line environment. It can also

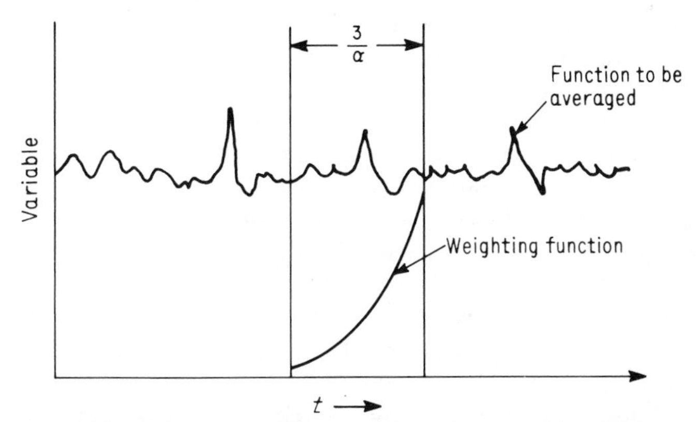

FIG. 7-3. Weighting function for the exponentially mapped average of a continuous process variable. (Reprinted by permission from Ref. 6.)

be safely assumed that information prior to time $3/\alpha$ in the past does not contribute significantly to the estimated averages. This can provide a guide to selecting α (or Q).

7-5 DISCRETE DESIGN METHODS

In the previous sections, the discussion began with a continuous model from which the discrete model was subsequently derived. Actually, considerable effort can be saved by simply postulating a discrete model initially. The most general form of a linear discrete model is as follows:

$$x_i = -\sum_{j=1}^{p} a_j x_{i-j} + \sum_{j=1}^{q} b_j u_{i-j} + D \tag{7-33}$$

In this form of the model, linear regression can always be used to calculate the coefficients. Dead time generally presents no problems, although one or more of the initial b-coefficients may be zero or nearly zero.

Since Eq. 7-33 is a difference equation, the corresponding pulse transfer function can be readily determined:

$$HG(z) = \frac{X(z)}{U(z)} = \frac{b_1 z^{-1} + \cdots + b_q z^{-q}}{1 + a_1 z^{-1} + \cdots + a_p z^{-p}} \tag{7-34}$$

This pulse transfer function can be used in any of the algorithm design methods presented in Chapter 6.

Kalman (7) describes a procedure by which the model in Eq. 7-33 can be incorporated into what he calls a self-optimizing control system, as illustrated in Fig. 7-4. The essential features of this approach are as follows:

1. From normal control operation, the sums of squares and cross products are calculated from the process input and output data. Exponentially mapped estimates or another type of weighting function is used so that old data is automatically descarded.
2. Since the model parameters can be calculated via linear regression, the computational effort is such that they can be updated frequently.
3. Although practically any algorithm design method can be used, that proposed by Kalman (presented in Sec. 6-6) is extremely attractive because of its simplicity.

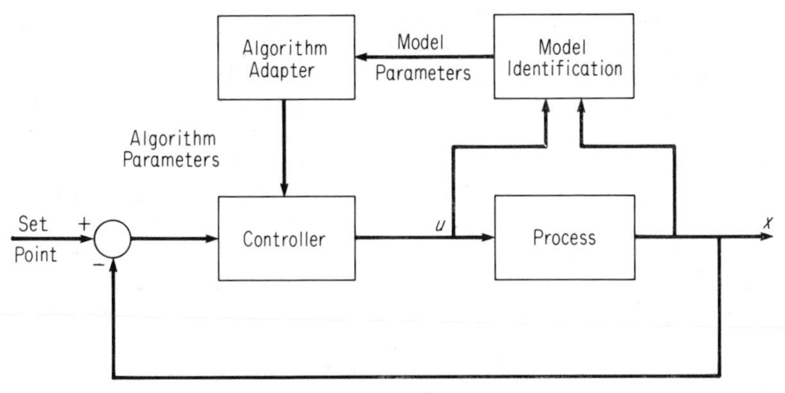

FIG. 7-4. Self-optimizing control system.

Although this procedure is quite simple, there are some problems. For example, if the process input and output remain at or near the same values for an extended period, the simultaneous equations encountered in the linear regression become linearly dependent and the model parameters can no longer be updated. But if control is so good, then there is little incentive to update the current model parameters.

Readers interested in pursuing this approach should consult the original article by Kalman for discussion of the points relevant to this procedure.

7-6 OTHER METHODS

This chapter has by no means presented all the modeling techniques currently available. Notable exceptions are those by Astrom (8), Koepcke (9), and Box and Jenkins (10–15). The Kalman filter has also been used in some applications (16–18) and should be attractive for a number of others.

LITERATURE CITED

1. Murrill, P. W., R. W. Pike, and C. L. Smith, "Dynamic Mathematical Models," *Chemical Engineering*, Sept. 9, 1968, pp. 117-120; Oct. 7, 1968, pp. 177-182; Nov. 18, 1968, pp. 165-169; Dec. 16, 1968, pp. 103-106; Jan. 27, 1969, pp. 167-172; Feb. 24, 1969, pp. 105-108; Mar. 10, 1969, pp. 111-116; Apr. 7, 1969, pp. 151-154; May 19, 1969, pp. 195-200; June 16, 1969, pp. 97-100; July 28, 1969, pp. 139-142; August 25, 1969, pp. 125-128.
2. Ritter, O. A., C. L. Smith, and P. W. Murrill, "Input Excitation Effects

on Experimental Model Development," presented at the 66th National A.I.Ch.E. Meeting, Portland, Oregon, Aug. 26, 1969.

3. Dahlin, E. B., "On-Line Identification of Process Dynamics", *IBM Journal of Research and Development*, Vol. 11, No. 4 (July 1967), pp. 406–426.
4. Moore, C. F., C. L. Smith, and P. W. Murrill, "Formulating the Nonlinear Least-Squares Model Regression for Fast On-Line Analysis," presented at the 64th National A.I.Ch.E. Meeting, New Orleans, March 16–20, 1969.
5. Wilde, D. J., *Optimum Seeking Methods*, Prentice-Hall, Englewood Cliffs, N.J., 1964.
6. "Continuous Data Analysis with Analog Computers Using Statistical and Regression Techniques," Bulletin No. ALAC 62023, Electronic Associates, Inc., 1964.
7. Kalman, R. E., "Design of Self-Optimizing Control System," *Transactions of the ASME*, February 1958, pp. 468-478.
8. Astrom, K. J., "Computer Control of a Paper Machine-Application of Linear Stochastic Control Theory," *IBM Journal of Research and Development*, Vol. 11, No. 4 (July 1967), pp. 389-405.
9. Koepcke, R. W., "A Discrete Design Method for Digital Control," *Control Engineering*, Vol. 13, No. 6 (June 1966), p. 83.
10. Box, G. E. P., and G. M. Jenkins, "Mathematical Models for Adaptive Control and Optimization", Technical Report No. 49, University of Wisconsin, (May, 1965).
11. ________, "Mathematical Models for Adaptive Control and Optimization," *A.I.Ch.E.-I Chem. Eng. Symposium Series No. 4*, (1965).
12. ________, "Some Statistical Aspects of Adaptive Optimization and Control", *J. R. Statist. Soc. B.*, 24 (1962), pp. 297–343.
13. ________, "Further Contributions to Adaptive Quality Control: Simultaneous Estimation of Dynamics: Non-Zero Costs," *Bulletin of the International Statistical Institute*, 34th Session, Ottawa, Canada (May, 1963).
14. Ramaker, B. L., C. L. Smith, and P. W. Murrill, "Development of Predictor Models," *Automatica*, Vol. 6, No. 2 (July 1970), p. 303.
15. ________, "Controlling a Noisy System," *Instrumentation Technology*, Vol. 17, No. 6 (June 1970), pp. 61–66.
16. Wells, C. H., "Application of Modern Estimation and Identification Techniques to Chemical Process," Joint Automatic Control Conference, Boulder, Colo., Aug. 5–7, 1969, pp. 473–481.
17. Wells, C. H., and R. E. Larson, "Combined Optimum Control and Estimation of Serial Systems with Time Delay," Joint Automatic Control Conference, Boulder, Colo., Aug. 5–7, 1969, pp. 23-33.
18. Wells, C. H., "Some Applications of Modern Control Theory to the Process Industry," Proceedings of the 25th Annual ISA Conference, Philadelphia, Oct. 26–29, 1970.

chapter 8

Advanced Control Techniques

The fact that a digital control system is more costly than a conventional analog control system is generally accepted today. In perhaps the majority of the cases, the justification for this extra capital outlay comes from improved control using advanced control techniques that are either impossible or impractical under the hardware constraints of the analog system. This chapter discusses several techniques in this category.

8-1 FEEDFORWARD CONTROL (1)

A feedforward control system is one that predicts the manipulative inputs which will keep the controlled variables at their desired values when measured disturbances enter the process. Therefore, feedforward control is normally a solution to the regulator problem (control of disturbances), but in many cases it can be used on the servomechanism (set point) problem. Feedforward control is typically compared to feedback control, which is based on the measurement of the controlled variables, the comparison of their measured values with their desired values, and the resultant usage of any differences between them as a means of manipulating inputs to eliminate these differences.

To summarize the difference between feedback and feedforward control when applied to the regulator problem, it should be noted that feedback control is based on measurement of the controlled variables, while feedforward control is based on measurement of disturbances. Because of this, they differ in when they use the manipulative inputs. Feedback control uses the manipulative inputs after the controlled variables have deviated from their desired values, and feedforward control uses the manipulative inputs before the controlled variables have deviated from their desired values. Figure 8-1 illustrates these ideas.

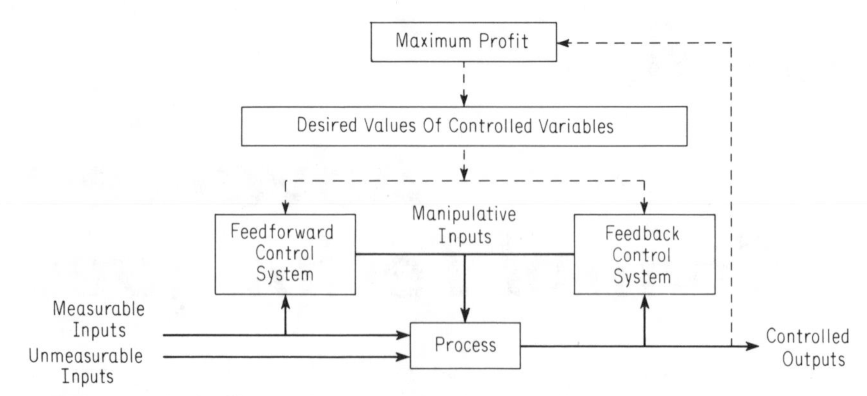

FIG. 8-1. A feedforward and feedback control system with solid lines indicating automatic information flow and dashed lines showing possible manual adjustment. (Reprinted with permission from J. A. Miller, P. W. Murrill, and C. L. Smith, "How to Apply Feedforward Control," *Hydrocarbon Processing*, July 1966, p. 165. Copyright 1967, by Gulf Publishing Company, Houston, Texas).

Why and when to use feedforward control can best be answered by further contrasting feedback and feedforward control. The concept and application of feedback control is relatively simple, but there are two main disadvantages to feedback control systems. In some systems these disadvantages are serious enough that feedback control may not be satisfactory.

Disadvantages of Feedback Control

One of these disadvantages is obvious—the control system does not and cannot take any corrective action to a system upset until after it has caused the controlled variable to deviate from its desired value. Feedforward control overcomes this disadvantage (if the upset can be measured) by taking control action before an error in the controlled variable occurs. (Theoretically, feedforward control systems are possible for which the controlled variables never deviate from their desired values, while in practice this is difficult to achieve.)

The second disadvantage of feedback control is really a result of the first—any corrective action that the control system takes is not felt until after the changing conditions have been propagated around the entire control loop, i.e., through each time lag in the loop. This means that not only must the controlled variable deviate from its desired value, but also, corrective action will lag in its effect because of signal propagation around the control loop. Feedforward control overcomes this disadvantage because corrective action is taken before the error occurs, and therefore, it is hoped that no error will exist for propagation around the loop.

Difficult Processes for Feedback Control

There are two process factors which typically make feedback control for the regulator problem unsatisfactory. These are the occurrence of frequent disturbances of large magnitude and/or large amounts of lag within the process. The terms frequent and large are not very exact because they really have to be defined on an economic basis, which is difficult to do even for a specific process. It is obvious why frequent disturbances may make feedback control unsatisfactory, and if these disturbances can be directly or indirectly measured, feedforward control can show improvement. The reason why large amounts of lag may make feedback control unsatisfactory goes back to the second disadvantage of feedback control because lags or dead times cause delay of signal corrective action.

Figure 8-2 quantitatively illustrates the effect of a large dead time or a large lag on the quality of feedback control. The three responses shown consist of a first-order-lag-plus-dead-time process with a conventional proportional-integral-derivative feedback controller. Process I has a large dead time of 1.0 min and process II has a small dead time of 0.1 minutes, while both processes have a 1.0-min time constant. Process III has a large lag of 10.0 min with θ/τ of 0.1 as in process I. To compare how well feedback control performs for each process, a unit step change in disturbance was made with each controller optimally tuned to minimize the integral of time multiplied by the absolute error (ITAE), i.e., $\int_0^\infty t|e(t)|\,dt$. Figure 8-2 shows that the controlled variables for the large dead time or large lag processes deviate much more or much longer from the desired value than the controlled variable for the small dead time and small lag process. Feedforward control of these three processes would be perfect if the disturbance could be perfectly measured and the process perfectly modeled. Most processes for which feedforward control is useful, such as distillation columns, paper machines, reactors, kilns, furnaces, etc., have large time lags and/or large dead times.

It is misleading to think that feedforward control can replace feedback control. The use of feedback control with feedforward control will be necessary for several reasons. First, the feedforward controller will be limited by the mathematical model and equipment

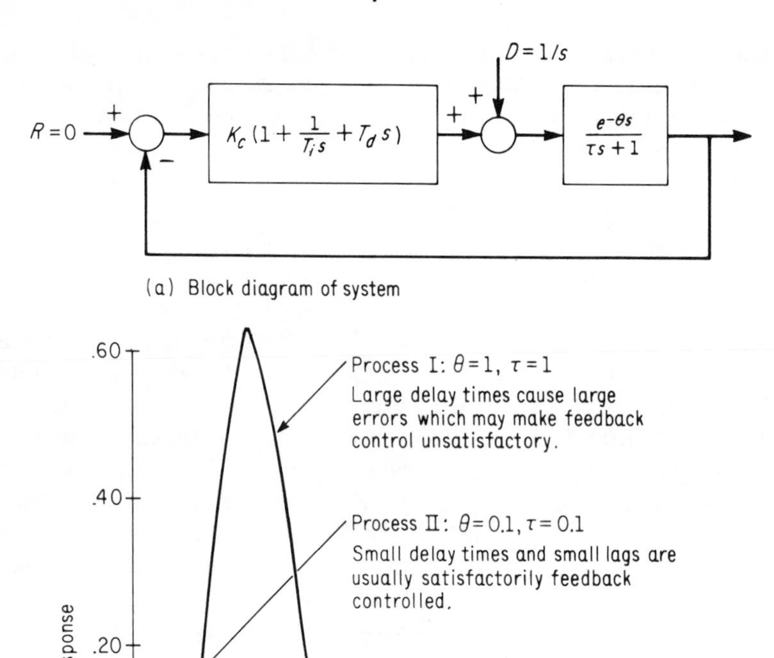

FIG. 8-2. Large delay times and/or large time lags are difficult to feedback control; however, they present no inherent problem for FFC.

accuracy. These inaccuracies may cause the feedforward controlled variable to have a steady-state offset, which can be eliminated with conventional feedback reset control. Second, the feedforward controller cannot compensate for unmeasured disturbances, while the feedback controller can. It is seen that feedforward and feedback control make a very natural combination, each compensating for the other's shortcomings.

While in theory combined feedforward/feedback control seems much more desirable than just feedback control, in practice it is often difficult to achieve. With feedforward control each measurable disturbance variable for which compensation is desired and its manipulated variable must be considered, and the change in the controlled variable that might result from them must be predictable in both a qualitative and quantitative manner. This implies that some type of control equipment must be associated with each of these measured disturbances, and the design engineer must thoroughly understand the effect on the controlled variable. These two facts make the application of feedforward control much more difficult than conventional feedback control, because the feedback system only requires one controller and any deviation in the controlled variable will initiate corrective action—whether or not the system or the designer knows why and how the deviation occurred. Despite the extra effort and equipment which feedforward control design requires, it can be very worthwhile economically. Much work has been done on feedforward control which will ultimately make it easier to apply.

Reference 1 includes a good literature survey on feedforward control. While there are some 42 references on feedforward control in this article, only five of them describe an actual application of feedforward control to plant-size equipment.

Steady-State Example

Although the application of feedforward control generally requires a process model of some type, the return from simple heat and material balances is often surprising, as will be illustrated for the heat exchanger in Fig. 8-3. A steady-state heat balance over the ex-

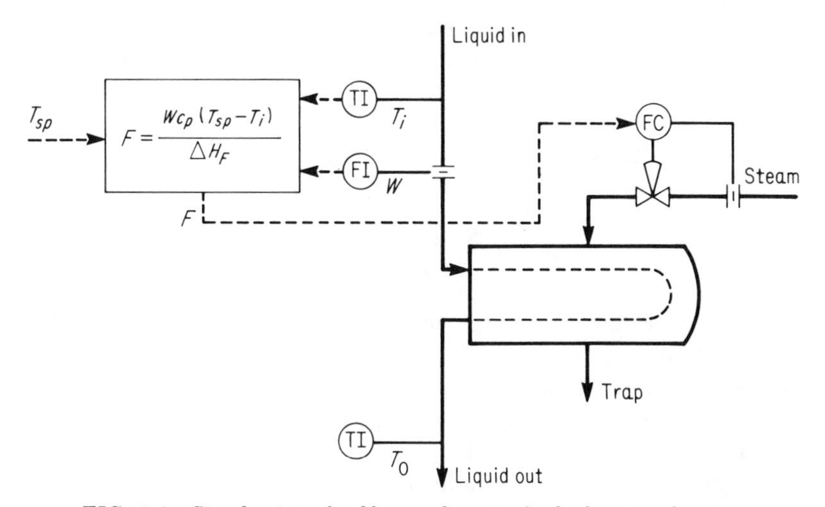

FIG. 8-3. Steady-state feedforward control of a heat exchanger.

changer gives the equation

$$Wc_p(T_0 - T_i) = F\Delta H_F \tag{8-1}$$

where
W = liquid flow, lb/hr
c_p = liquid heat capacity, Btu/lb °F
T_i = liquid inlet temperature, °F
T_0 = liquid outlet temperature, °F
F = steam flow, lb/hr
ΔH_F = heat released by steam, Btu/lb

This equation can be solved for F.

$$F = \frac{Wc_p(T_0 - T_i)}{\Delta H_F} \tag{8-2}$$

Replacing T_0 by the set point T_{sp} gives an equation that can be solved for the steam flow to give the desired outlet temperature.

$$F = \frac{Wc_p(T_{sp} - T_i)}{\Delta H_F} \tag{8-3}$$

As illustrated in Fig. 8-3, the flow W and inlet temperature T_i are measured. Reasonable values for c_p and ΔH_F should be readily available in most cases.

Note specifically from Fig. 8-3 that the controlled variable, the outlet temperature T_0, is not utilized by the feedforward controller in any manner. While the feedforward controller utilizing Eq. 8-3 should produce values of T_0 near the set point T_{sp}, slight discrepancies could arise from several sources. For example, the heat loss to the ambient surroundings was neglected in deriving Eq. 8-1. Also, slight errors may exist in the numerical values of c_p and ΔH_F.

Feedback control is ideally suited for eliminating the discrepancies between T_0 and T_{sp}. In this case, there are at least three ways in which this can be implemented.

1. A constant (or fudge factor) K_F could be added to Eq. 8-3 to give

$$F = \frac{Wc_p(T_{sp} - T_i)}{\Delta H_F} + K_F$$

   The output of the feedback controller would be K_F, as illustrated in Fig. 8-4a.
2. The output of the feedback controller could be the set point to the feedforward controller, as illustrated in Fig. 8-4b.
3. The output of the feedback controller could be the numerical value of one of the parameters in the feedforward controller as illustrated in Fig. 8-4c.

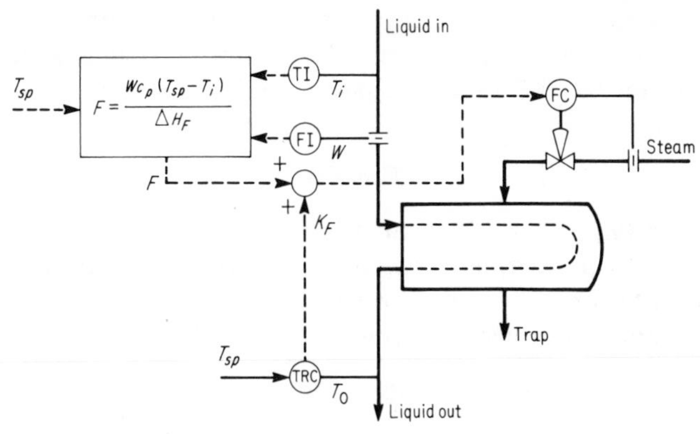

FIG. 8-4a. Feedback controller used to bias output of feedforward controller.

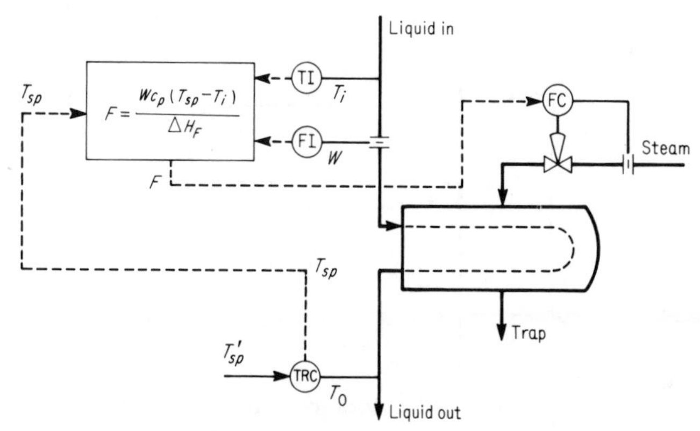

FIG. 8-4b. Output of feedback controller used to adjust the set point of the feedforward controller.

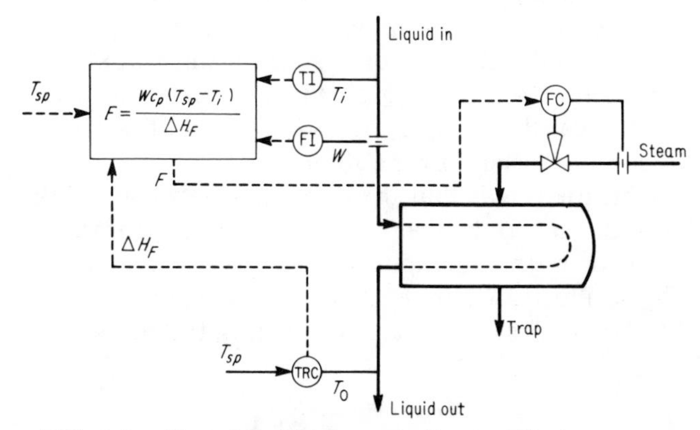

FIG. 8-4c. Use of feedback controller to adjust a parameter in the feedforward controller.

The scheme in Fig. 8-4c is quite similar to a commonly accepted procedure with ratio controllers. For example, if it is assumed that T_{sp} and T_i in Eq. 8-3 seldom change significantly, this equation reduces to

$$F = KW \tag{8-4}$$

where $K = c_p(T_{sp} - T_i)/\Delta H_F$ and the feedforward control system reduces to a ratio controller. A feedback controller is frequently used to adjust the ratio constant K of the ratio controller, as illustrated in Fig. 8-5. Analog hardware for the task is readily available.

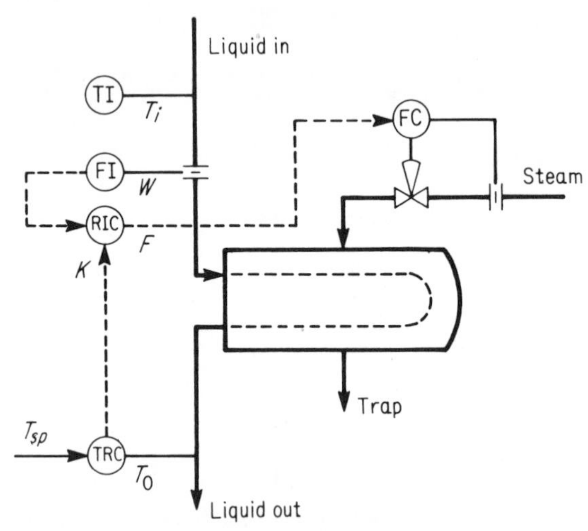

FIG. 8-5. Ratio controller.

Dynamic Compensation

Figure 8-6 illustrates the effectiveness of feedforward control as applied to a heat exchanger similar to the one in Fig. 8-3. The performance of a feedback control system is illustrated in Fig. 8-6a. The improvement from using a steady-state feedforward controller such as Eq. 8-3 can be seen in Fig. 8-6b. While the improvement is significant, some transient error still occurs.

The reason for this is that the dynamics between the disturbance (W or T_i) and the controlled variable (T_0) are different from the dynamics between the manipulated variable F and the controlled variable T_0, as illustrated in Fig. 8-7. For example, suppose both $G_1(s)$ and $G_2(s)$ are first-order lags (neglecting the process gains).

$$G_1(s) = -\frac{1}{\tau_1 s + 1}$$

$$G_2(s) = \frac{1}{\tau_2 s + 1}$$

Suppose a step change in W occurs and the step change in F necessary to exactly offset this change in W at steady-state (Eq. 8-3) is made

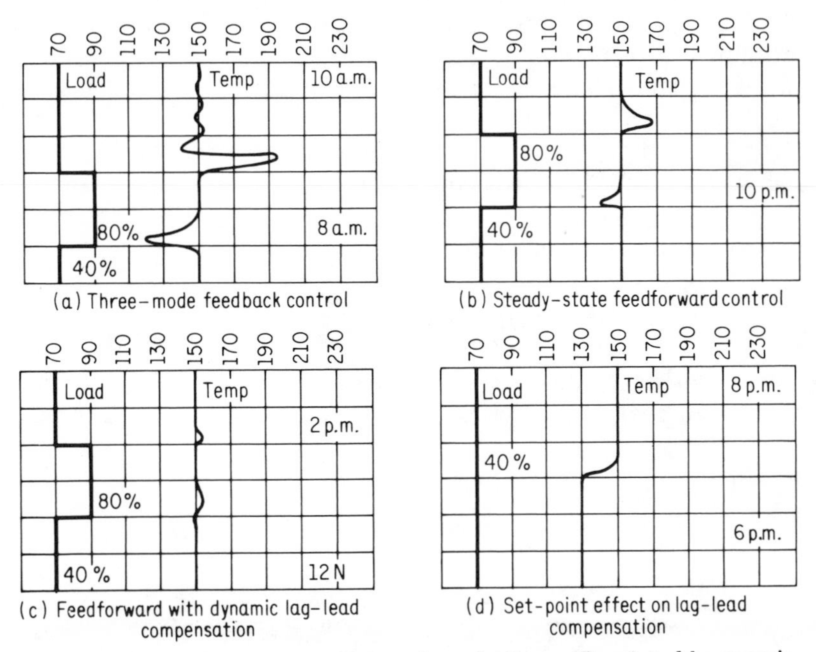

FIG. 8-6. Performance of feedforward controllers. (Reprinted by permission from Ref. 3.)

immediately. If $\tau_1 < \tau_2$, the response in T_0 would be as illustrated in the type of behavior that occurs in Fig. 8-6b.

For the idealized case in Fig. 8-7, the transient offset can be eliminated by incorporating the lead-lag element

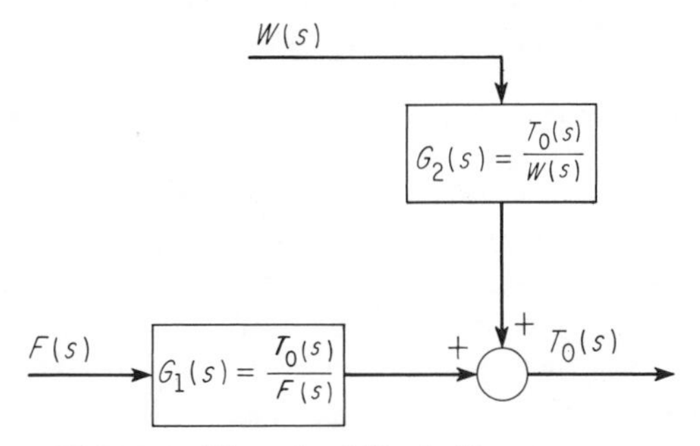

FIG. 8-7. Linearized block diagram representation of the heat exchanger.

$$G(s) = \frac{\tau_2 s + 1}{\tau_1 s + 1}$$

into the feedforward controller. While this element will not give perfect compensation for practical systems, it will generally improve the performance of the system to where little further improvement can be obtained, as is illustrated in Fig. 8-6c for the heat exchanger. In some cases, addition of a time delay to the dynamic compensation also helps.

The dynamic compensation is generally inserted at the output of the steady-state feedforward controller. Selection of the parameters τ_1 and τ_2 in the lead-lag element is generally done on-line. Shinskey (2) discusses this problem at length.

8-2 CASCADE CONTROL SYSTEMS (4)

For sake of discussion, suppose we have a process essentially consisting of two lags in series, as illustrated in Fig. 8-8. The usual

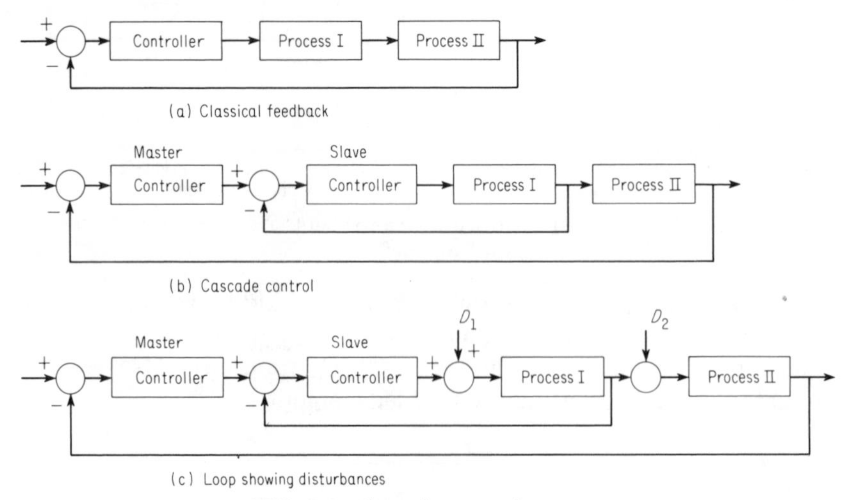

FIG. 8-8. Cascade control system.

feedback arrangement would be to measure the output and adjust the input as illustrated in Fig. 8-8a.

Assuming the dynamics of process I are somewhat faster than those of process II, the performance of this control system can be improved by somehow making process I even faster dynamically. One approach by which this can be accomplished is to add a feedback loop around process I, giving the cascade arrangement in Fig. 8-8. The output of the outer or master controller is the set point to the inner or slave controller.

This approach to the development of the cascade control concept was chosen in an attempt to show that cascading improves the basic performance of the loop. The real payout comes when the disturbances are considered, as illustrated in Fig. 8-8c. Some disturbances enter the inner loop, and others enter the outer loop. While cascading improves the compensation of disturbances entering the outer loop to some extent, compensation of those entering the inner loop is greatly improved. Utopia is when all major disturbances enter a fast inner loop.

Figure 8-9 illustrates a practical example of a cascade control system for a jacketed reactor.

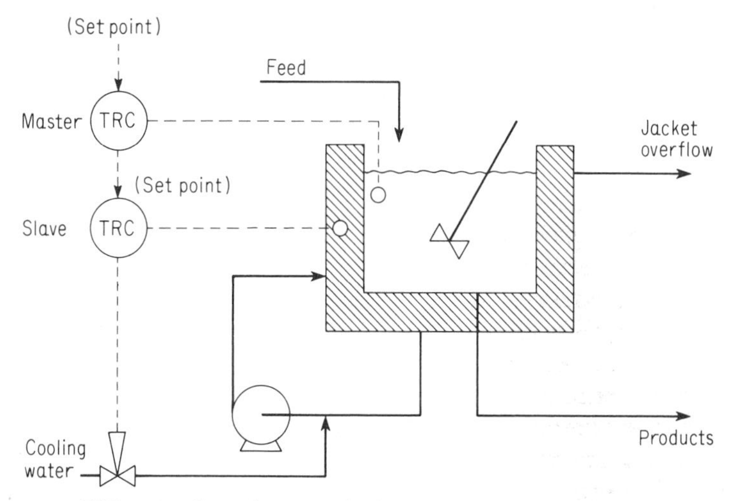

FIG. 8-9. Cascade control of a stirred jacketed reactor.

The main obstacle to obtaining good performance from a cascade control arrangement is the tuning of the two controllers. The basic approach is simple (5):

1. Place the master controller on manual and tune the inner loop using the method of your choice. As the inner loop must respond well to the set point changes from the master controller, it should probably be tuned to give good set-point response.
2. Place the inner loop on automatic and tune the outer loop. It should be noted that the "process," i.e., the entire entity controlled by the master controller, contains the secondary controller. It is suggested that the settings for the secondary controller not be changed while tuning the master controller.

Tuning is certainly a factor to consider in the selection of the modes for the various controllers. For the inner loop, a pure proportional controller is frequently recommended as offset is not significant for the intermediate variable (jacket temperature in the example in Fig. 8-9). Reset is sometimes added (especially in flow loops, where the proportional action would be too responsive to noise), but rarely is derivative found to be beneficial. For the outer loop, a PI controller is typically used, with rate added in some cases.

For a cascade system to function properly, the inner loop must be at least as fast as the outer loop, and preferably faster. If this condition is not met, it is generally impossible to satisfactorily tune the outer controller. Use of pure proportional action in the inner loop makes the inner loop faster, which should lead to an easier tuning problem.

For conventional analog control, the extra hardware required is a transducer for the intermediate variable, a second controller, and a remote set-point station. For digital control systems, the only extra hardware investment is the transducer. This is not to imply, however, that both loops must be digital. Very frequently the digital master loop provides the set point to a cascaded flow controller. Nor is it necessary that the master loop be feedback only.

8-3 MULTIVARIABLE CONTROL SYSTEMS

Up to this point we have considered only individual control loops designed to maintain one variable (the controlled variable c) at a desired value (the set point) by adjusting one process input (the manipulated variable m). Many practical control systems do not exactly fit this description, an example being a control system for the reactor in Fig. 8-10 that is designed to keep the reactor temperature and concentration near their desired values by adjusting both the rate of heat addition q and the reactor feed rate w. This system has two manipulated variables and two controlled variables, and is thus a multivariable control system.

The steady-state reactor temperature T is a function of the rate of heat addition q and reactor feed rate w, as represented by the functional relationship

$$T = f_1(q,w)$$

Similarly, the steady-state reactor concentration C_A is also a function of the manipulated variables as represented by

$$C_A = f_2(q,w)$$

Expressing these two equations as changes about an operating point gives

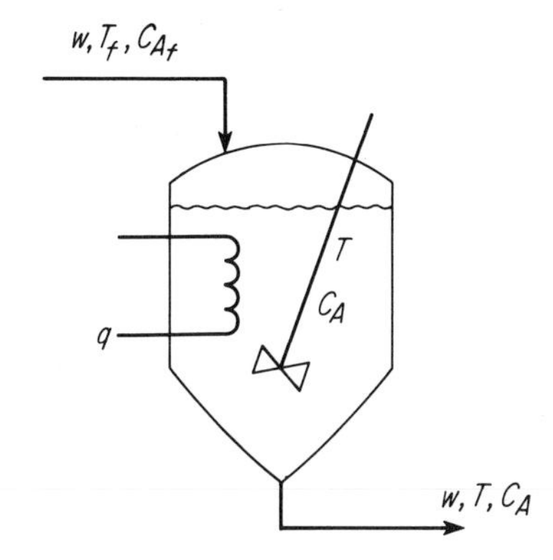

w = reactor feed rate, lb/hr

T_f = feed temperature, °F

C_{A_f} = feed concentration, lb-mole/ft^3

T = reactor temperature, °F

C_A = reactor concentration, lb-mole/ft^3

V = reactor volume, ft^3

q = rate of heat addition, Btu/hr

ΔH = heat of reaction, Btu/mole A consumed

c_p = heat capacity of reacting mass, Btu/lb °F

ρ = density of reacting mass, lb/ft^3

Reaction: $2A \longrightarrow B$

Rate of disappearance of A (lb-mole/hr): $k_0 e^{-a/T} C_A^2$

FIG. 8-10. Chemical reactor.

$$\Delta T = \left.\frac{\partial T}{\partial q}\right|_w \Delta q + \left.\frac{\partial T}{\partial w}\right|_q \Delta w \tag{8-5}$$

$$\Delta C_A = \left.\frac{\partial C_A}{\partial q}\right|_w \Delta q + \left.\frac{\partial C_A}{\partial w}\right|_q \Delta w \tag{8-6}$$

The matrix representation of these equations is

$$\begin{bmatrix} \Delta T \\ \\ \Delta C_A \end{bmatrix} = \begin{bmatrix} \left.\dfrac{\partial T}{\partial q}\right|_w & \left.\dfrac{\partial T}{\partial w}\right|_q \\ \left.\dfrac{\partial C_A}{\partial q}\right|_w & \left.\dfrac{\partial C_A}{\partial w}\right|_q \end{bmatrix} \begin{bmatrix} \Delta q \\ \\ \Delta w \end{bmatrix} \tag{8-7}$$

or in matrix form,

$$\mathbf{c} = \mathbf{M}\mathbf{m} \tag{8-8}$$

where the i, jth element of M, is $\left.\dfrac{\partial c_i}{\partial m_j}\right|_m$. This is typically represented schematically as in Fig. 8-11.

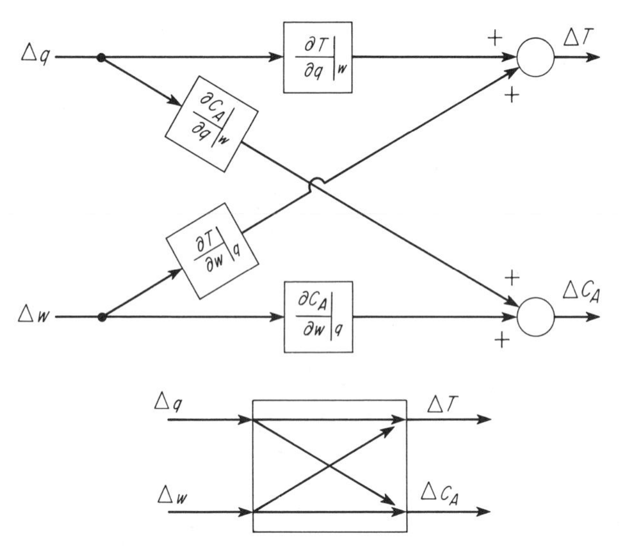

FIG. 8-11. Typical representations of a multivariable control system.

The matrix of the partials in Eq. 8-7 is called the *gain matrix*, each element being equivalent to the process gain in a single-input, single-output control loop. We see for the two-input, two-output system there are four "gains"; for an n-input, n-output system there will be n^2 of them. These "gains" could be determined experimentally with n process tests, e.g., impose a step change in one manipulated variable while holding all others constant. The response of each controlled variable could be recorded for evaluation of the partial of each controlled variable with respect to the manipulated variable for which the step change was made.

That is, suppose we place all control systems on the reactor on manual and impose a change Δw on the flow rate into the reactor while holding q constant. The resulting steady-state change in C_A and T, say ΔC_A and ΔT, can be obtained from the response of the unit. Since q has been held constant the partials with respect to w, can be evaluated as follows:

$$\left.\frac{\partial C_A}{\partial w}\right|_q \cong \frac{\Delta C_A}{\Delta w} \tag{8-9}$$

$$\left.\frac{\partial T}{\partial w}\right|_q \cong \frac{\Delta T}{\Delta w} \tag{8-10}$$

The "approximately equal to" notation is necessary because most systems are nonlinear which means that the final answers will vary somewhat with the magnitude of Δw.

However, the above "gains" are not the only ones that can be defined (2). For example, suppose a step change in the rate of heat addition is made in the configuration in Fig. 8-12. As the control

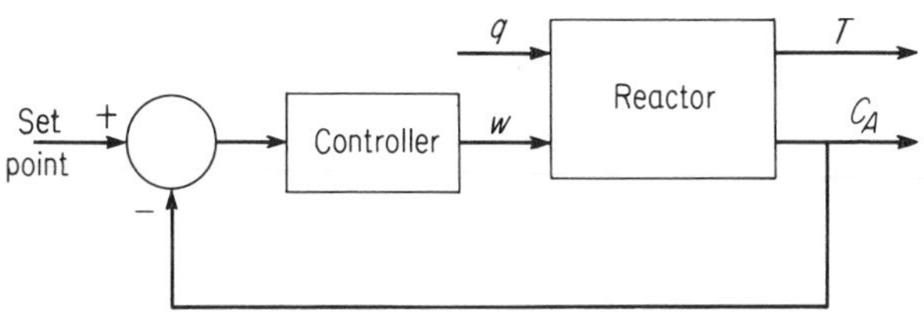

FIG. 8-12. Configuration to evaluate $\left.\frac{\partial T}{\partial q}\right|_{C_A}$

system will vary the reactor feed rate w to maintain constant concentration C_A, the change in temperature T due to a change in q will not be given by $\left.\frac{\partial T}{\partial q}\right|_w$, as developed above. Instead, this test can be used to evaluate another gain, namely $\left.\frac{\partial T}{\partial q}\right|_{C_A}$, the change in temperature with respect to jacket flow rate at constant reactor concentration.

Obviously one can define three additional gains of this type, namely, $\left.\frac{\partial T}{\partial w}\right|_{C_A}$, $\left.\frac{\partial C_A}{\partial q}\right|_T$, and $\left.\frac{\partial C_A}{\partial w}\right|_T$. The general notation for these will be $\left.\frac{\partial c_i}{\partial m_j}\right|_c$.

In principle, experimental tests analogous to that illustrated in Fig. 8-10 could be used but such tests are not really very convenient. Suppose then that we undertake to calculate the gains $\left.\frac{\partial c_i}{\partial m_j}\right|_c$ from the gains $\left.\frac{\partial c_i}{\partial m_j}\right|_m$. For the specific gain $\left.\frac{\partial T}{\partial w}\right|_{C_A}$, this is initialized from Eq. 8-7 by setting ΔC_A to zero.

$$\begin{bmatrix} \Delta T \\ 0 \end{bmatrix} = \begin{bmatrix} \left.\frac{\partial T}{\partial q}\right|_w & \left.\frac{\partial T}{\partial w}\right|_q \\ \left.\frac{\partial C_A}{\partial q}\right|_w & \left.\frac{\partial C_A}{\partial w}\right|_q \end{bmatrix} \begin{bmatrix} \Delta m \\ \Delta w \end{bmatrix} \tag{8-11}$$

This effectively gives two equations, which can be combined to give

$$\left.\frac{\partial T}{\partial w}\right|_{C_A} \cong \left.\frac{\Delta T}{\Delta w}\right|_{C_A} = \left.\frac{\partial T}{\partial w}\right|_{q} - \left\{\left.\frac{\partial T}{\partial q}\right|_{w} \cdot \left.\frac{\partial C_A}{\partial w}\right|_{q} \Big/ \left.\frac{\partial C_A}{\partial q}\right|_{w}\right\} \tag{8-12}$$

A similar approach can be used to obtain the other gains $\left.\frac{\partial c_i}{\partial m_j}\right|_c$. This is also equivalent to the relationship

$$\mathbf{C} = (\mathbf{M}^{-1})^T \tag{8-13}$$

where the element on the ith row and jth column of **C** is the reciprocal of $\left.\frac{\partial c_j}{\partial m_j}\right|_c$.

These two definitions of the open-loop gains can be used to define a relative gain as follows (2):

$$\lambda_{ij} = \frac{\partial c_i/\partial m_j|_m}{\partial c_i/\partial m_j|_c} \tag{8-14}$$

It should be noted that the relative gain is simply the product of the corresponding elements in matrices **C** and **M**.

The utility of the relative gain is that it indicates how manipulated and controlled variables should be paired. For example, should the rate of heat addition to the reactor in Fig. 8-9 be used to control reactor temperature or reactor concentration? In some cases the pairing is obvious, but in some cases it is not. Occasionally it makes no difference (equal coupling), while it is also possible for the proper pairing to change with operating level. In these difficult cases, the relative gains can be indispensible.

To facilitate the pairing of manipulated variables to controlled variables, it is convenient to arrange the relative gains as follows:

	m_1	m_2	$\cdots$	m_n
c_1	λ_{11}	λ_{12}	$\cdots$	λ_{1n}
c_2	λ_{21}	λ_{22}	$\cdots$	λ_{2n}
.	.	.	.	.
.	.	.	.	.
.	.	.	.	.
c_n	λ_{n1}	λ_{n2}	$\cdots$	λ_{nn}

The pairing is via the combinations whose relative gains are the largest positive values. For example, the manipulated variable for c_1 would be the one corresponding to the largest positive relative gain on the first row of the relative gain matrix.

A useful property of the relative gain matrix is that each row and each column sums to 1. Thus in a 2×2 matrix as for the reactor in Fig. 8-9, only one relative gain must be determined.

Even though the above analysis indicates the proper pairing of controlled and manipulated variables, connecting simple feedback controllers often proves to be unsatisfactory. In tuning the controllers, the general approach is to first tune one of the loops and then tune the other. Generally, no problems arise in tuning the first controller, giving the configuration in Fig. 8-13. In this configura-

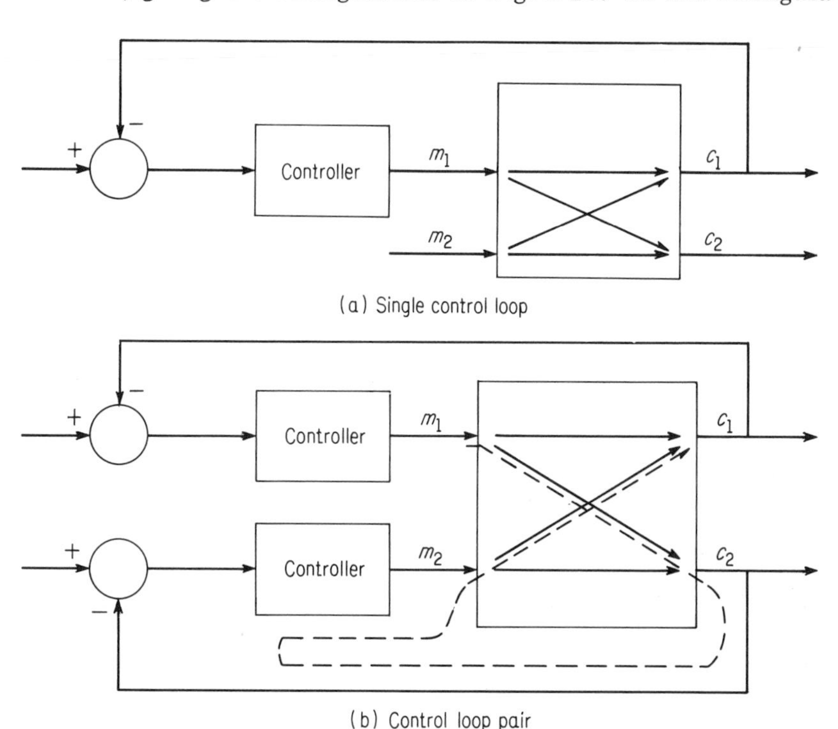

FIG. 8-13. Conventional approach to control multivariable systems.

tion, the gain of the system is $\left.\dfrac{\partial c_1}{\partial m_1}\right|_{m_2}$. Placing the second controller on automatic gives the control system in Fig. 8-13b. The gain of the top loop is now different due to the feedback via the lower loop. The value of the gain is now $\left.\dfrac{\partial c_1}{\partial m_1}\right|_{c_2}$. If this value is significantly different from the original gain for which the top loop was tuned, its performance deteriorates significantly. These loops are then said to be *interacting*.

As an example of interaction, suppose the rate of heat flow controls the temperature and the reactor feed rate controls the concentration in the reactor in Fig. 8-9. If the concentration of the reactor feed increases, the composition controller would decrease the reactor feed rate. This would tend to raise the reactor temperature (feed temperature is below reactor temperature), calling for a decrease in the rate of heat addition. But this temperature rise would increase the reaction rate, calling for more feed to maintain the composition. Thus a control action in one loop is a disturbance to the other.

To circumvent this problem, a decoupler can be used. The process-decoupler combination should be designed such that the two control loops appear to be independent of one another, i.e., $\left.\frac{\partial T}{\partial m_2}\right|_{m_1} = \left.\frac{\partial C_A}{\partial m_1}\right|_{m_2} = 0$. Consider the representation of the decoupler and process shown in Fig. 8-14. Suppose we require that a unit

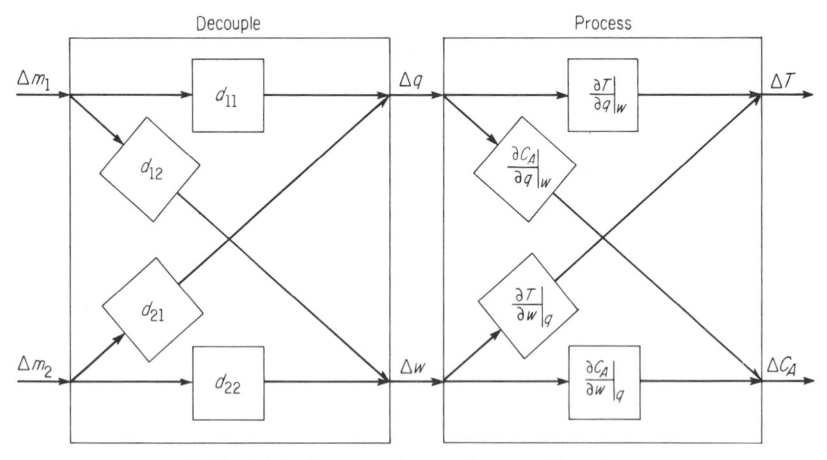

FIG. 8-14. Process-decoupler combination.

change in m_1 produce a unit change in T but no change in C_A. Similarly, a unit change in m_2 should produce a unit change in C_A but no change in T. Under these conditions, the equation describing the process-decoupler combination in Fig. 8-14 becomes

$$\begin{bmatrix} \Delta T \\ \Delta C_A \end{bmatrix} = \begin{bmatrix} 1 & 0 \\ 0 & 1 \end{bmatrix} \begin{bmatrix} \Delta m_1 \\ \Delta m_2 \end{bmatrix} \tag{8-15}$$

$$\begin{bmatrix} 1 & 0 \\ 0 & 1 \end{bmatrix} = \begin{bmatrix} \left.\frac{\partial T}{\partial q}\right|_w & \left.\frac{\partial T}{\partial w}\right|_q \\ \left.\frac{\partial C_A}{\partial q}\right|_w & \left.\frac{\partial C_A}{\partial w}\right|_q \end{bmatrix} \begin{bmatrix} d_{11} & d_{12} \\ d_{21} & d_{22} \end{bmatrix} \tag{8-16}$$

or

$$\begin{bmatrix} d_{11} & d_{12} \\ d_{21} & d_{22} \end{bmatrix} = \begin{bmatrix} \left.\dfrac{\partial T}{\partial q}\right|_w & \left.\dfrac{\partial T}{\partial w}\right|_q \\ \left.\dfrac{\partial C_A}{\partial q}\right|_w & \left.\dfrac{\partial C_A}{\partial w}\right|_q \end{bmatrix}^{-1} \tag{8-17}$$

That is, the decoupler matrix should be the inverse of the process gain matrix.

Any time the inverse of a matrix appears, some attention should be devoted to the question of when this inverse exists. Mathematically, this means that the determinant of the matrix must be nonzero. If the controlled variables have been chosen carefully, this will generally be the case. However, if one of the controlled variables is functionally related to the other by a function that does not contain both manipulated variables, then the matrix is likely to be singular.

In the above approach, all four gains of the decoupler-process combination were specified. In reality the only ones that are really necessary are the following:

$$\left.\frac{\partial T}{\partial m_2}\right|_{m_1} = \left.\frac{\partial C_A}{\partial m_1}\right|_{m_2} = 0 \tag{8-18}$$

Denoting the remaining gains by k_1 and k_2, Eq. 8-15 becomes

$$\begin{bmatrix} \Delta T \\ \Delta C_A \end{bmatrix} \begin{bmatrix} k_1 & 0 \\ 0 & k_2 \end{bmatrix} \begin{bmatrix} \Delta m_1 \\ \Delta m_2 \end{bmatrix} \tag{8-19}$$

Equation 8-17 now becomes

$$\begin{bmatrix} d_{11} & d_{12} \\ d_{21} & d_{22} \end{bmatrix} = \begin{bmatrix} \left.\dfrac{\partial T}{\partial q}\right|_w & \left.\dfrac{\partial T}{\partial w}\right|_q \\ \left.\dfrac{\partial C_A}{\partial q}\right|_w & \left.\dfrac{\partial C_A}{\partial w}\right|_q \end{bmatrix}^{-1} \begin{bmatrix} k_1 & 0 \\ 0 & k_2 \end{bmatrix} \tag{8-20}$$

Suppose we now require that d_{11} and d_{22} equal one. Equation 8-10 is now essentially four equations in four unknowns, namely d_{12}, d_{21}, k_1, and k_2. Solving for d_{12} and d_{21} gives the decoupler illustrated in Fig. 8-15.

A closer examination of Fig. 8-15 reveals that, for all practical purposes, this decoupler is essentially two feedforward elements. For example, suppose a change Δm_1 occurs in m_1. Then q would be changed the same amount. Due to the coupling in the process, a change in q is effectively a disturbance to the bottom loop. Therefore, we feed the change in q (or m_1) forward through the gain d_{12} and change w. If d_{12} is selected properly, the change in C_A due to

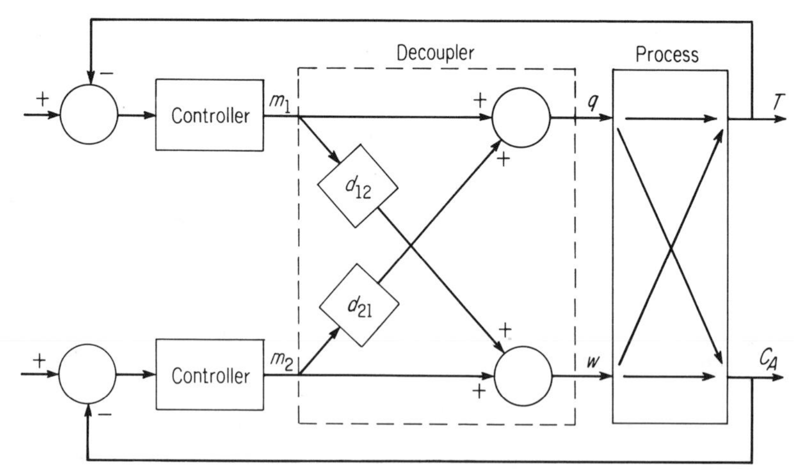

FIG. 8-15. A decoupler consisting essentially of two feedforward elements.

the change in q should be exactly offset by the change in C_A due to the change in w.

While a steady-state decoupler, i.e., one consisting of gains only, will generally produce significant improvement over no decoupling at all, addition of dynamics to the decoupler will frequently give significant further improvement, and in some cases is necessary so that the performance of the decoupler is satisfactory. Since we have shown that the decoupler in Fig. 8-15 essentially consists of two feedforward elements, it seems reasonable to follow the customary practice for feedforward systems and place a lead-lag compensator in elements d_{12} and d_{21}.

As an example, suppose the process consists of four first-order lags with unity gains:

$$G_{11}(s) = \frac{1}{\tau_{11}s + 1} \qquad G_{12}(s) = \frac{1}{\tau_{12}s + 1}$$

$$G_{21}(s) = \frac{1}{\tau_{21}s + 1} \qquad G_{22}(s) = \frac{1}{\tau_{22}s + 1}$$

The process can be represented by the following matrix notation:

$$\begin{bmatrix} C_1(s) \\ C_2(s) \end{bmatrix} = \begin{bmatrix} G_{11}(s) & G_{12}(s) \\ G_{21}(s) & G_{22}(s) \end{bmatrix} \begin{bmatrix} M_1(s) \\ M_2(s) \end{bmatrix}$$

where M denotes an input to the decoupler.

For complete decoupling, the process-decoupler should be represented by the following matrix:

$$\begin{bmatrix} K_1(s) & 0 \\ 0 & K_2(s) \end{bmatrix}$$

This matrix must equal the product of the following matrices:

$$\begin{bmatrix} G_{11}(s) & G_{12}(s) \\ G_{21}(s) & G_{22}(s) \end{bmatrix} \begin{bmatrix} 1 & D_{12}(s) \\ D_{21}(s) & 1 \end{bmatrix}$$

This equality produces the following four equations:

$$G_{11}(s) + G_{12}(s)D_{21}(s) = K_1(s)$$
$$G_{21}(s) + G_{22}(s)D_{21}(s) = 0$$
$$G_{11}(s)D_{12}(s) + G_{12}(s) = 0$$
$$G_{21}(s)D_{12}(s) + G_{22}(s) = K_2(s)$$

The second and third equations give the solutions for $D_{21}(s)$ and $D_{12}(s)$.

$$D_{21}(s) = -\frac{G_{21}(s)}{G_{22}(s)} = -\frac{\tau_{22}s + 1}{\tau_{21}s + 1}$$

$$D_{12}(s) = -\frac{G_{12}(s)}{G_{11}(s)} = -\frac{\tau_{11}s + 1}{\tau_{12}s + 1}$$

Note that both of these are lead-lags.

8-4 ADAPTIVE CONTROL AND ON-LINE TUNING

The typical process plant possesses two characteristics that degrade the performance of the customarily used linear control laws. First, these plants are nonlinear; second, they are nonstationary (i.e., their characteristics change with time). The natural suggestion for such plants is to use an adaptive control system. For the past fifteen years or so several books and articles have appeared on the general theory and philosophy of such systems (7-12). Not only has the number of resulting practical applications been disappointing, but these efforts have also failed to produce a consistent definition of what constitutes an adaptive control system.

As the term "adaptive" has come to be used in process control systems, the definition by Truxal is perhaps most appropriate (9). By his definition, an adaptive system would automatically compensate for variations in system dynamics by adjusting the controller characteristics so that the overall system performance would be satisfactory. Such a system would include elements to measure or estimate the process dynamics, and other elements to change the controller characteristics accordingly.

Gibson (44), on the other hand, does not advocate such a broad definition, insisting that the adaptive system measure the index of performance. Under his definition, most of the techniques presented later in this section would be termed "nonlinear control systems" rather than "adaptive."

The so-called model-reference adaptive control system illustrated in Fig. 8-16 is probably the most commonly considered adaptive

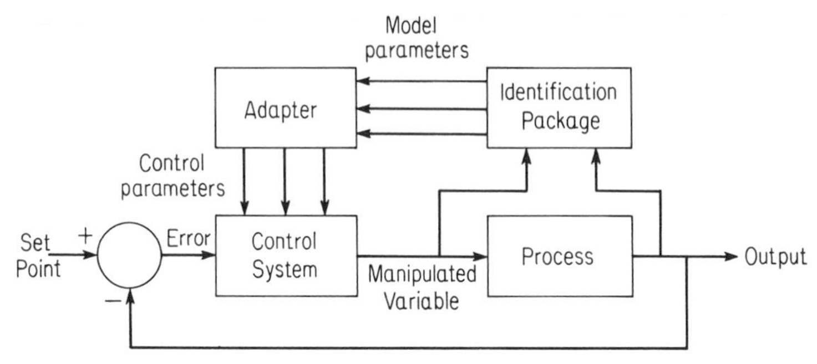

FIG. 8-16. Model-reference adaptive control system.

system. A basic requirement is an identification procedure applicable to the process under control. In order to simplify the identification facet of the problem, the model is often a rather simple linear model. For the control law being used, the proper values of the parameters to give the desired control system performance are computed from the model and its parameters.

As the first example, consider the on-line tuning approach of Gallier and Otto (13). They suggest that the process be periodically pulsed from the computer as illustrated in Fig. 8-17, and the resulting process output recorded. Using Marquardt's nonlinear regression program (14), the parameters in the following second-order-lag-plus-dead-time model are determined.

$$\frac{X(s)}{M(s)} = \frac{Ke^{-\theta s}}{(\tau_1 s + 1)(\tau_2 s + 1)}$$

In order to save time, it is suggested that this equation be transformed into its difference-equation counterpart in much the same manner as suggested in Sec. 7-3. Finally, the proportional gain, reset time, and derivative time (if this mode is present) must be related to the model parameters K, θ, τ_1, and τ_2 in the model. To accomplish this last step, Callier and Otto present tuning maps.

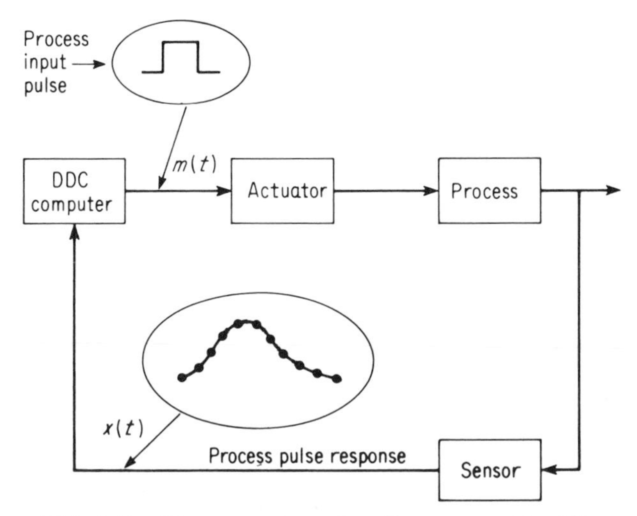

FIG. 8-17. The computer pulses the process, identifies characteristic process coefficients, and tunes the control algorithm according to some arbitrary criterion. (Reprinted by permission from Ref. 13.)

Being a common plague to adaptive strategies, one drawback of the approach presented above is the test that must be made on the process. Due to the pulse intentionally imposed on the input, the system deviates from the desired operating point, generally resulting in a hopefully brief period of less efficient operation which must be justified by the resulting improvement in control system performance from tuning.

In a few rare cases, it is possible to measure or infer the value of a parameter such as the gain of the system. One such case is the nonlinear gain of valves (15). If the characteristics of the valve are known, the gain of the control valve can be calculated from the stem position. Neglecting the change in valve dynamics with operating level, the adaptive policy is to change the gain of the control algorithm in such a manner that the product of the valve gain and controller gain remains constant. This procedure has become known as *adaptive gain tuning*.

For situations in which test signals are not permissible and process parameters cannot be inferred directly from process measurements, methods based on a concept originally due to Marx (16, 17) may be applicable. The basic control loop with Marx's frequency servo superimposed is illustrated in Fig. 8-18 for adaptive gain tuning. The objective is to maintain the product of the controller gain K_c and the process gain K at some desired value. Although this approach

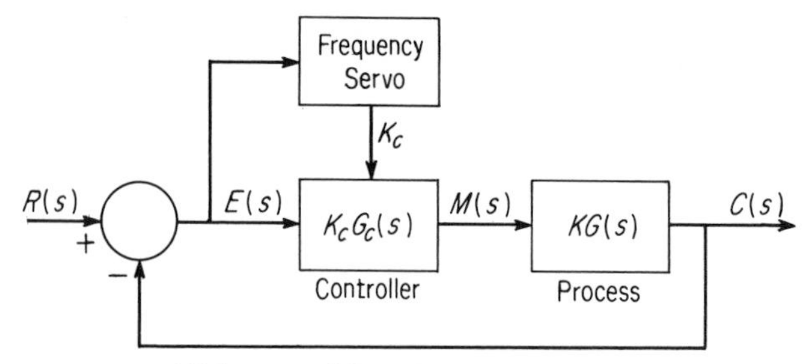

FIG. 8-18. Marx frequency servo.

can be applied to adapt parameters other than the controller gain, we shall only illustrate this case.

This method relies heavily on the frequency characteristics of the error signal. For example, consider the frequency spectra of the error signal (produced by a step change in set point) for different values of the loop gain KK_c illustrated in Fig. 8-19. Note that as the

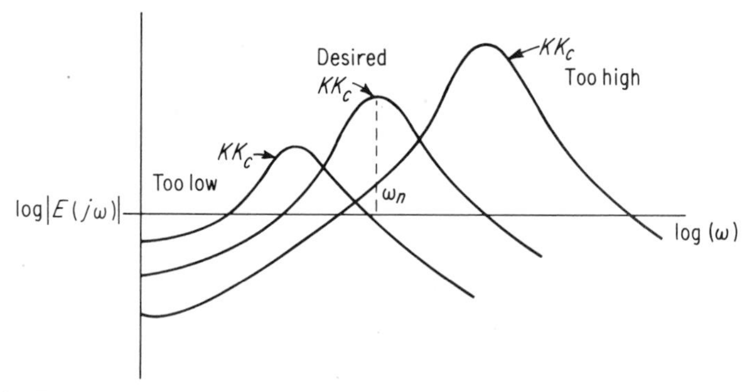

FIG. 8-19. Error spectra for various values of the loop gain KK_c.

gain increases, the amplitude at low values of ω decreases and the amplitude at high values of ω increases, i.e., they are gain-sensitive. The frequency servo illustrated in Fig. 8-20 uses a high-pass filter followed by a full-wave rectifier (absolute value) to detect changes in the high-frequency region. Similarly, a low-pass filter followed by a full-wave rectifier detects changes in the low-frequency region. Since the amplitudes at high frequencies equals that at low frequencies for the desired value of KK_c in Fig. 8-19, the outputs of the two rectifiers are compared and the difference integrated to change the controller gain. When the gain is too low, the rectified output of the low-pass filter exceeds that of the high-pass filter, giving a positive error which, upon integrating, increases the controller gains. The converse is true when the gain is too high.

Before proceeding with other modifications of this procedure, a

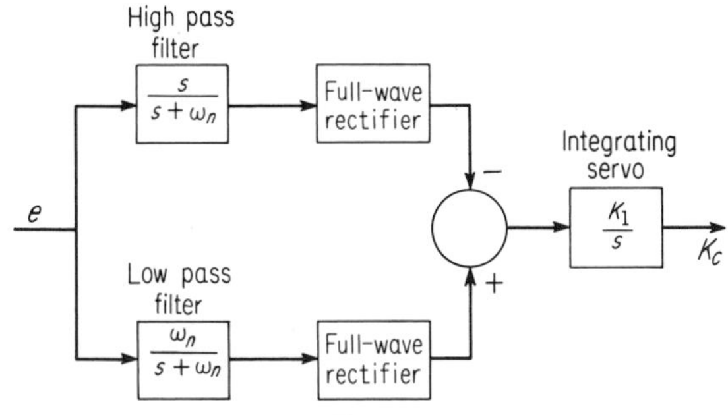

FIG. 8-20. Frequency servo.

word is in order concerning how the plots in Fig. 8-19 are obtained. If $G_c(s)$ and $G(s)$ are known, the error is given by

$$E(s) = \frac{R(s)}{1 + KK_c G_c(s) G(s)}$$

and the frequency spectrum can be plotted for any chosen input $R(s)$, such as a step input. For the typical case in which $G(s)$ is not known, the selected input can be made to the process and the error signal recorded for various values of the controller gain K_c. Performing a frequency analysis on this data gives the plot in Fig. 8-19.

Certain improvements in this approach have been suggested by various workers (17, 18):

1. The high- and low-pass filters can be replaced by bandpass filters to give greater selectivity.
2. Weighting functions can be incorporated into the comparator to give more generality.
3. To obtain the controller gain, the output of the comparator could be the input to the usual three-mode algorithm instead of the simple integrator of Fig. 8-20.

The resulting configuration is illustrated in Fig. 8-21. An even further modification is replacing the full-wave rectifiers by squarers followed by low-pass filters (19).

This approach is not without its problems. The frequency spectra in Fig. 8-19 upon which the frequency servo is designed are for a specific input; stability problems have been noted for systems subjected to other inputs or subjected to noise (19).

8-5 DEAD-TIME COMPENSATION

A common characteristic of process systems is the presence of a significant dead time, which Shinskey (2) has chosen to appropriately label the "difficult element to control." The purpose of this section

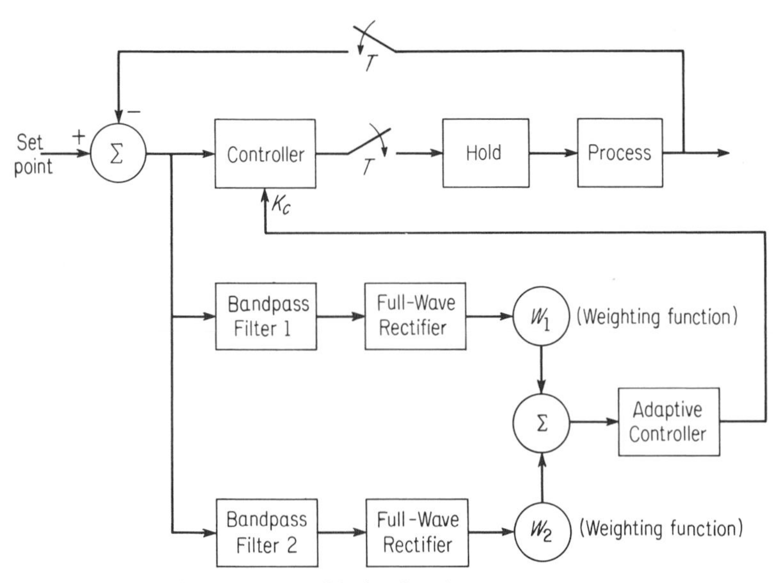

FIG. 8-21. Bakke's adaptive gain tuning.

is to examine some special control techniques whose performance exceeds that of the basic feedback system for loops with dead time.

Suppose the plant transfer function is composed of a dead time θ and a transfer function $G(s)$ containing no dead time. The typical feedback-control arrangement for this plant is illustrated in Fig. 8-22a. Note that the feedback variable is the output of the dead time element. However, if the feedback variable could be the output of $G(s)$, the dead time is effectively moved outside the control loop, thereby allowing the controller to be tuned more tightly and provide better control. The practical difficulty with this concept is that $G(s)$ and the dead time θ do not occur as distinct elements in the real plant but are distributed throughout. The signal cannot therefore be measured directly. However, the basic principle of dead-time compensation is to provide an approach whereby this signal can effectively be measured.

Suppose by some means we construct a process model composed of a dead time θ_m and a transfer function $G_m(s)$ containing no dead time. Since the output of the controller in Fig. 8-22a can be measured, suppose it is passed through the model and used to cancel the original feedback signal as illustrated in Fig. 8-22b. If the model were perfect, i.e., if $G_m(s) = G(s)$ and $\theta_m = \theta$, the output from the summer would be zero and there would be no feedback to the controller in Fig. 8-22b. For the realistic case of imperfect modeling, the output from the summer is the model error.

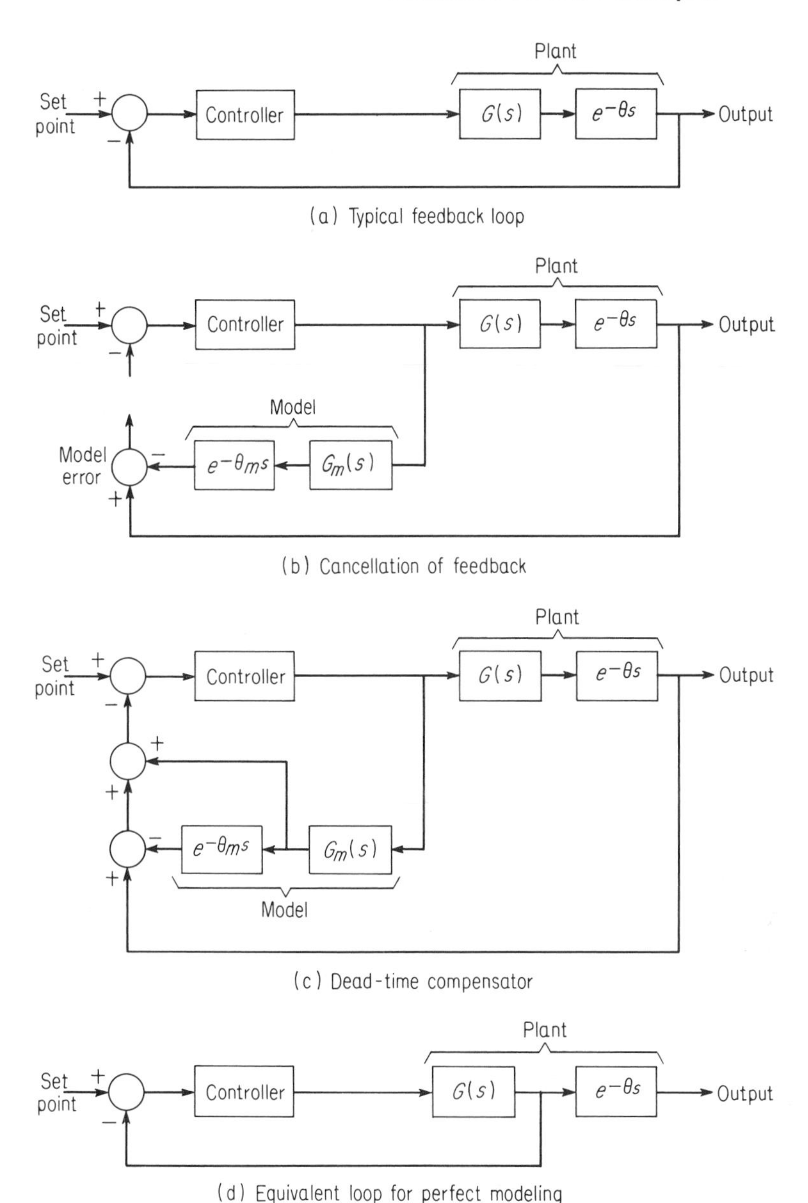

FIG. 8-22. Development of dead-time compensator.

As pointed out earlier, the most desirable feedback would be the output from the transfer function $G(s)$. Although this signal is not measurable, the output signal from the model transfer function $G_m(s)$ is available and can be used for feedback. Adding this to the block diagram in Fig. 8-22b gives the control system in Fig. 8-22c.

This control system was first proposed by O. J. M. Smith (21), and has come to be known as the *Smith predictor* or *dead-time compensator*. From Fig. 8-22d note that the effective result for perfect modeling is to move the dead time outside the feedback loop.

Although Smith originally proposed this strategy for analog control systems, the difficulty of implementing the model dead time using analog circuits prevented his idea from gaining wide acceptance. But with the advent of digital computer control, the dead time could be readily implemented. Bakke (22) advocated the utility of this approach, and showed that significant improvements could be gained using the simple first-order-lag-plus-dead-time model. He also showed that it was practical to implement this strategy for variable time delays.

The performance of dead-time compensation on a practical system is illustrated in Fig. 8-23. These responses are for the tempera-

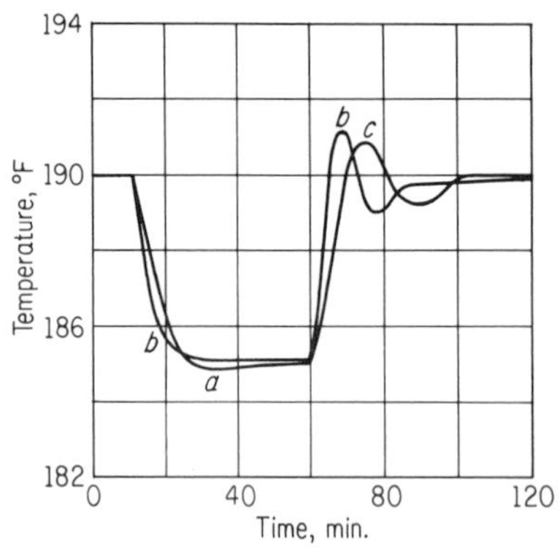

(a) Performance of dead-time compensation for changes in set point; PI control; ITAE criterion tuned to set point changes; sampling time = 5 min; *a*, regular feedback; *b*, dead-time compensation.

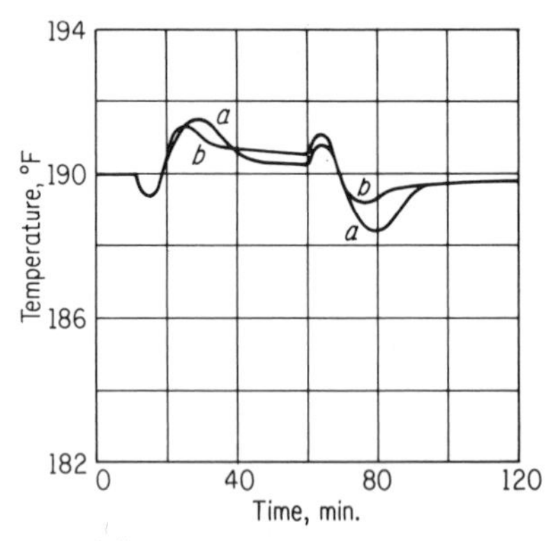

(b) Performance of dead-time compensation for changes in reactor feed rate; PI control; ITAE criterion tuned to set point changes; sampling time = 5 min; *a*, regular feedback; *b*, dead-time compesation.

FIG. 8-23. Performance of dead-time compensation. (Reprinted by permission from Ref. 23.)

ture control of a stirred, jacketed chemical reactor in which a second-order, temperature-dependent exothermic reaction is proceeding (23). Figure 8-23 is for set-point changes; Fig. 8-23b is for changes in a disturbance (the reactor feed rate). The model used was

$$G_m(s) = \frac{Ke^{-\theta s}}{\tau s + 1}$$

where $K = -0.0103°F/\text{min}$
$\theta = 2.3 \text{ min}$
$\tau = 12.5 \text{ min}$

Even more improvement over regular feedback control would be obtained for systems with a larger dead time to time constant ratio.

Dead-time compensation should be considered only when using conventional PI or PID control algorithms. Algorithms designed by z-transform methods inherently contain dead-time compensation.

The predictor as proposed by Smith is essentially free-running, i.e., the controller output is continuously fed to the model to generate the model output. A potential disadvantage of this method is that model errors accumulate with time. Wheater (24) indicates that the predictor is considerably more sensitive to the accuracy of the model dead time than is the conventional PI controller. One potential avenue to circumvent this problem when implementing digitally is to construct a predictor from the model $G_m(s)$ as illustrated in Fig. 8-24.

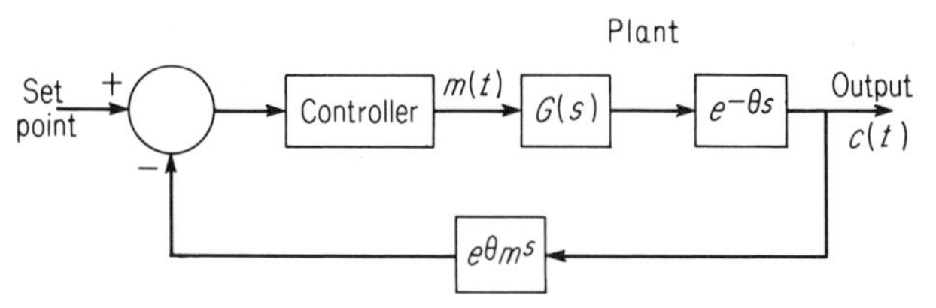

FIG. 8-24. Use of a predictor directly.

The predictor simply uses the current value of the process output c_o and the values of the controller output $u(t)$ for the past θ_m units of time to estimate the process output θ_m units of time in the future. If $G(s)$ is a first-order lag, namely $K/(\tau s + 1)$, the predictor solves the differential equation

$$\tau \frac{dc(t)}{dt} + c(t) = m(t)$$

$$c(t_0) = c_o$$

where t_0 is the current time and c_0 is the current output. The analytic solution is

$$c(t_0 + \theta_m) = c_o e^{-t_0/\tau} + K \int_{t_0}^{t_0+\theta_m} u(\lambda - \theta_m) e^{-(t_0-\lambda)/\tau} \, d\lambda$$

The integral could be evaluated numerically, or alternatively, the original differential equation could be solved numerically from time t_0 to time $t_0 + \theta_m$. For digital systems, a difference equation could also be developed (see exercises).

Moore (25) suggests that, for digital systems, the predictor be used to estimate $c(t_0 + \theta_m + T/2)$, thereby compensating for the phase shift in the interface in addition to the process dead time. His controller is proportional plus integral, but the integral term is based on the model error instead of the control error. This gives consistent performance for both load and set-point changes, thereby circumventing the problem described in Sec. 6-9. This modification could also be incorporated into the regular dead-time compensation.

LITERATURE CITED

1. Miller, J. A., P. W. Murrill, and C. L. Smith, "How to Apply Feedforward Control," *Hydrocarbon Processing*, Vol. 48, No. 7 (July 1969), pp. 165–172.
2. Shinskey, F. G., *Process Control Systems*, McGraw-Hill, New York, 1967.
3. ——, "Feedforward Control Applied," *ISA Journal*, Vol. 10, No. 11 (November 1963), p. 61.
4. Wills, D. M., "Cascade Control Applications and Hardware," Technical Bulletin Number TX119-1, Minneapolis Honeywell Regulator Co., Philadelphia, 1960.
5. Murrill, P. W., *Automatic Control of Processes*, Intext Educational Publishers, Scranton, Pa., 1967.
6. Dahlin, E. B., et al., "Designing and Tuning Digital Controllers," *Instruments and Control Systems*, Vol. 41, No. 7 (July 1968), p. 87.
7. Aseltin, J. A., A. R. Mancini, and C. W. Sarture, "A Survey of Adaptive Control Systems," *IRE Trans. on Automatic Control*, December 1958.
8. Gibson, J. E., "Making Sense Out of the Adaptive Principle," *Control Engineering*, Vol. 8, No. 8 (August 1961).
9. Mishkin, E. and L. Braun, *Adaptive Control Systems*, McGraw-Hill, New York, 1961, pp. 4–6.
10. Sklansky, J., "Adaptation and Feedback," AIEE (JACC Paper 16–1), June 1962.
11. Bellman, R. E., *Adaptive Control Processes, A Guided Tour*, Princeton U.P., 1961.
12. Eveleigh, V. W., *Adaptive Control and Optimization Techniques*, McGraw-Hill, New York, 1968.
13. Gallier, P. W., and R. E. Otto, "Self-Tuning Computer Adapts DDC Algorithms," *Instrumentation Technology*, Vol. 15, No. 2 (February 1968), p. 65.
14. Marquardt, D. W., "An Algorithm for Least-Squares Estimation of Control Parameters," *Journal of Society of Industrial and Applied Mathematics*, June 1963.
15. Pemberton, T. J., "Automatic Tuning of Nonlinear Control Loops," *Instruments and Control Systems*, Vol. 41, No. 5 (May 1968), p. 123.
16. Marx, M. F., "Recent Adaptive Control Work at the General Electric Company," *Proceedings of the Self-Adaptive Flight Controls Symposium*, January 1959.
17. Mishkin, E., and L. Braun, *Adaptive Control Systems*, McGraw-Hill, New York, 1961, pp. 327–332.

18. Bakke, R. M., "Adaptive Gain Tuning Applied to Process Control," Paper presented at 19th Annual Instrument Society of America Conference and Exhibit, Oct. 12–15, 1964, New York. (ISA preprint 3.2-1-64.)
19. Keyes, M. A., and J. A. Gudaz, "Industrial Applications of Adaptive Control," *Instruments and Control Systems*, Vol. 43, No. 7 (July 1970) pp. 93–96.
21. Smith, O. J. M., "Close Control of Loops with Dead Time," *Chemical Engineering Progress*, Vol. 53, No. 5 (May 1967), pp. 217–219.
22. Bakke, R. M., "Direct Digital Control with Self-Adjustment for Processes with Variable Dead Time and/or Multiple Delays," 20th Annual ISA Conference and Exhibit, Los Angeles, October 1965.
23. Corripio, A. B., and C. L. Smith, "Computer Simulation to Evaluate Control Strategies," *Instruments and Control Systems*, Vol. 44, No. 1 (January 1971).
24. Moore, C. F., "Improved Algorithm for Direct Digital Control," *Instruments and Control Systems*, Vol. 43, No. 1 (January 1970), pp. 70–74.

chapter 9

Optimal Control

With the advent of the digital computer, one anticipated result was the widespread application of optimal control theory to processes. Nothing could be further from the fact at present. Economical applications of optimal control theory are few and far between. Why then devote a chapter to the subject? Because current efforts are identifying requirements unique to process units, and with a thorough understanding and consideration of these, progress may eventually be made toward successful applications of optimal control theory to process units.

The purpose of this chapter is not to derive the various theorems of optimal control theory, which are well developed in several books devoted to the subject (1,2,3). Instead, the main thrust will be to point out the requirements and characteristics unique to processes, and to present potential avenues to take them into consideration.

9-1 THE CASE FOR FEEDBACK

One absolutely essential requirement of process control systems is that they be feedback in nature, a property which we shall see is not always met by optimal control formulations. This requirement will be illustrated by an example, which will not only illustrate the case for feedback but also introduce the concept of *dynamic programming* (to be discussed in more detail in a subsequent section of this chapter).

Consider a logistics problem in which a traveler wishes to pass from state 1 to any one of states 4, 5, or 6, as illustrated in Fig. 9-1a. No matter how he proceeds, he must pass through either state 2 or state 3. Suppose at each state he has the perogative of calling either "up" or "down" to select the alternate paths. The cost of traversing

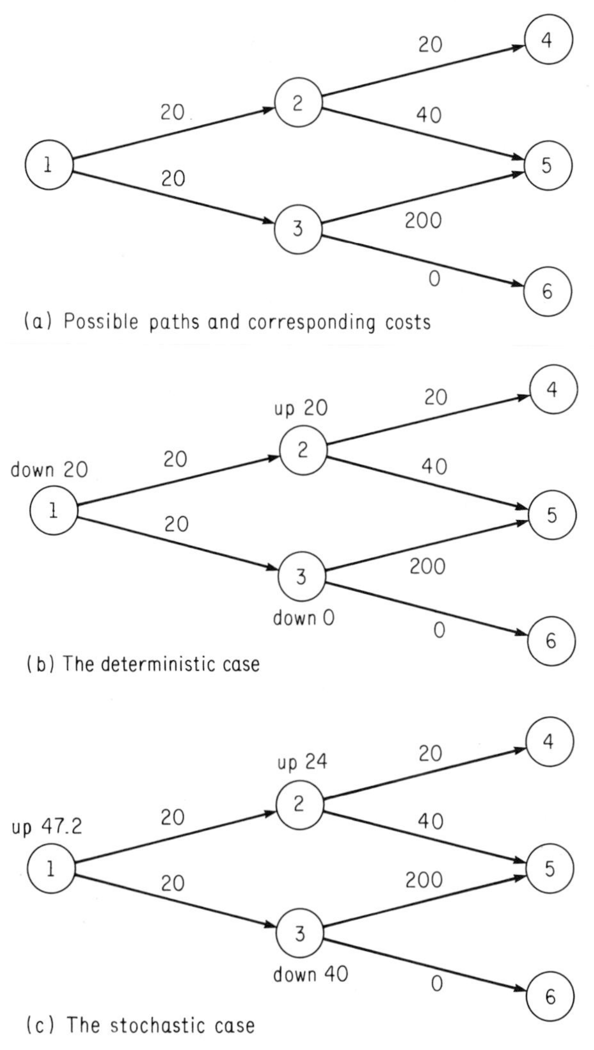

FIG. 9-1. Logistics problem.

each path is as shown in Fig. 9-1a. The objective is to select the path with the least cost for the trip.

The dynamic programming approach to such problems is to break them into smaller problems. For example, suppose for the moment that our traveler has arrived at state 2. We ask the question, "What should he do now?" As calling "up" costs 20 units and "down" costs 40, he obviously calls "up." It is convenient to write the decision and the cost beside the state, as illustrated in Fig. 9-1b. From state 3, the obvious decision is to call "down" for no cost.

Now we can proceed to state 1. If he calls "up," the trip to state 2 costs 20 units. As we already know that the minimum cost from state 2 to his ultimate destination is 20 units, the total trip via state 2 costs 40 units. However, if he calls "down," he can attain state 3 for a cost of 20 units, from whence he can reach his destination at no further cost. Thus his decision at state 1 is to call "down" to reach state 3, from which he should again call "down" to reach his destination.

So far this problem has been completely deterministic. But this is really unlike our usual process that is subject to so many random disturbances and changes. Suppose we introduce some uncertainty into our problem by supposing that our traveler does not always get his wishes. Specifically, if he calls "up" he will actually get his wish only 80 percent of the time, getting "down" the other 20 percent. Calling "down" gives similar results; i.e., he gets "down" only 80 percent of the time and "up" the other 20 percent.

The analysis of the problem proceeds as before. If he has attained state 2, the following options and corresponding expected costs are open to him:

1. Call "up." Cost = $(.80)(20) + (.20)(40) = 24$
2. Call "down." Cost = $(.80)(40) + (.20)(20) = 36$

Thus he calls "up," anticipating a cost of 24 units "on the average," as illustrated in Fig. 9-1c. At state 3, the following options are open:

1. Call "up." Cost = $(.80)(200) = 160$
2. Call "down." Cost = $(.20)(200) = 40$

Thus he calls "down." At state 1, the possibilities are

1. Call "up." Cost = $.80\,(20 + 24) + .20\,(20 + 40) = 47.2$
2. Call "down." Cost = $.80\,(20 + 40) + (.20)(20 + 24) = 56.8$

Consequently he should call "up."

Obviously his initial decision is different for the two cases. However, it is his next decision in which we are interested. As calling "up" does not assure him of traveling in this direction, he must also consider the other alternative. If he indeed goes "up," his best call at the intermediate state 2 is "up." However, if he goes "down" from state 1, he should call "down" at state 3. The point is that his call at the intermediate state depends upon the state at which he is located. This decision cannot be made at state 1, since it is not known which state will be encountered next.

This is effectively the same reason that we must use feedback for process control. The control decision at a given point in time must be based upon an examination of the state of the process at that time, not on how it got there. This is the basic concept of feedback.

For this reason we shall in this chapter concentrate our attention toward techniques that yield a feedback control law.

9-2 THE INTEGRAL MODE

Perhaps as many as 90 percent of the conventional analog process controllers in use contain the integral mode. Students in courses in basic automatic control theory learn that the primary reason for this is to eliminate offset, the steady-state difference between the set point and the process output. Let us reexamine the function of the integral mode to see why it is so necessary in many process applications. Suppose the process is at steady-state with some value of the manipulated variable (the input to the process) and a corresponding value of the process output. Now suppose it is desired to change the process output to a new value. This means that a new steady-state value of the manipulated variable must somehow be obtained.

Basically, there are two avenues to accomplish this. First, it may be accomplished via trial-and-error tests on the process, i.e., try a value of the manipulated variable and see if the corresponding process output equals the desired value. If not, the value of the manipulated variable is changed again. The function of the integral mode in the process controller is to implement this procedure. The input to the integral mode is the process error, the difference between actual process output and the desired output. Based on this input, the integrator changes the value of the manipulated variable.

The second avenue is to calculate from a process model the value of the manipulated variable to give the desired process output. This implementation is effectively used by optimal control theory. Most problem formulations are to transfer the system (or process) from some initial condition to the desired equilibrium point, (i.e., desired process output and corresponding value of the manipulated variable), which is assumed to be known beforehand. Whenever this last assumption can be met, the integral mode is indeed unnecessary.

For process systems, this last assumption rarely holds. First, process models seldom are sufficiently accurate, especially when linearized approximations are used. Second, the process is subjected to disturbing inputs in addition to the manipulated variable. Whenever these disturbances cannot be measured (the majority of the cases), then it is impossible to calculate the value of manipulated variable to give the desired process output at the equilibrium point.

This too has been an obstacle to the application of optimal control theory to process systems. In addition to being feedback in nature, practical control strategies for many processes must in some manner incorporate the integral mode.

9-3 THE STATE EQUATION

To illustrate the concepts to be discussed in this section, consider the classical "black box" problem in which an unknown system can be subjected to one or more time varying inputs. We shall call the collection of the m inputs $u_1(t), u_2(t), \ldots, u_m(t)$ the input vector $\mathbf{u}(t)$. Suppose the black box has k outputs $y_1(t), y_2(t), \ldots, y_k(t)$ which we shall call the output vector $\mathbf{y}(t)$. There is also a set of n variables $x_1(t), x_2(t), \ldots, x_n(t)$ that completely define the state of our system. The collection of these will be called the state vector $\mathbf{x}(t)$. The system may also have some adjustable parameters (e.g., valves open or closed, steam pressure, variable resistors, etc.) that are not normally manipulated inputs, but these we shall ignore for the moment.

Suppose the system is initially at some state $\mathbf{x}_0$ at time t_0. If over the interval $t_0 < t \leqslant t_1$ the input to the system is $\mathbf{u}(t)$, the state $\mathbf{x}(t_1)$ can be mathematically represented by the relationship:

$$\mathbf{x}(t_1) = f_1\{\mathbf{x}_0, \mathbf{u}(t)\} \tag{9-1}$$

This says that the state at time t_1 is completely determined by the initial state at time t_0 and the input or control action taken over the interval $t_0 < t \leqslant t_1$. Such systems are said to be *dynamical*, as contrasted to systems for which the state $\mathbf{x}(t_1)$ depends only upon $\mathbf{u}(t_1)$, the input at time t_1.

We normally like to represent our dynamical physical systems by differential equations. In this light, we like to think of our systems as generating a succession of states in response to the inputs. The differential equation relates the change in state, i.e., the derivative $d\mathbf{x}(t)/dt$ or $\dot{\mathbf{x}}(t)$, to the present state $\mathbf{x}(t)$ and the present input $\mathbf{u}(t)$. Mathematically, this becomes

$$\dot{\mathbf{x}}(t) = \mathbf{f}\{\mathbf{x}(t), \mathbf{u}(t)\} \tag{9-2}$$

This equation embodies all the information in Eq. 9-1 except the initial state. Thus, we need an initial condition, namely

$$\mathbf{x}(t_0) = \mathbf{x}_0 \tag{9-3}$$

This allows the state $\mathbf{x}(t_1)$ to be determined from Eq. 9-2 for the specified input $\mathbf{u}(t)$ over the interval $t_0 < t \leqslant t_1$.

Equation 9-2 is typically called the state equation of the system. The output $\mathbf{y}(t_1)$ is related to the state by the equation

$$\mathbf{y}(t_1) = \mathbf{g}\{\mathbf{x}(t_1),\mathbf{u}(t_1)\} \tag{9-4}$$

That is, the output at any instant of time is completely specified by the state and the inputs at that instant of time.

To illustrate these concepts, suppose we consider the description of the reactor in Fig. 9-2. The reactor is a well-mixed, continuous

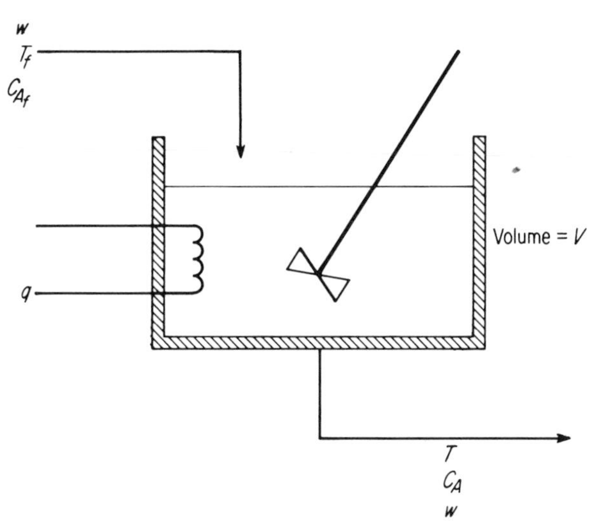

FIG. 9-2. Stirred chemical reactor.

flow unit in which the second-order reaction

$$2A \longrightarrow B$$

occurs. For simplicity, the rate constant k is assumed to be independent of temperature. The heat of reaction ΔH is based on one mole of A consumed. Making a heat and material balance over the reactor gives the following equations:

$$\frac{dT}{dt} = \frac{w}{V\rho}(T_f - T) + \frac{q}{V\rho c_p} - \frac{(\Delta H)kC_A{}^2}{\rho c_p} \tag{9-5}$$

$$\frac{dC_A}{dt} = \frac{w}{V\rho}(C_{A_f} - C_A) - kC_A{}^2 \tag{9-6}$$

The state variables for this system would be the reactor temperature T and concentration C_A, i.e.,

$$\mathbf{x}(t) = \begin{bmatrix} T \\ C_A \end{bmatrix} \tag{9-7}$$

As manipulated inputs, the reactor feed rate w and rate of heat input q are logical selections. Thus the vector $\mathbf{u}(t)$ is

$$\mathbf{u}(t) = \begin{bmatrix} q \\ w \end{bmatrix} \tag{9-8}$$

The functions f_1 and f_2 in the state Eq. 9-2 become the right-hand sides of Eqs. 9-5 and 9-6.

These equations are of course nonlinear. As usual, a linear set would be much more convenient. Such equations can be obtained by linearizing Eqs. 9-5 and 9-6 about an equilibrium point $\overline{T}, \overline{C}_A, \overline{w}$, and $\overline{q}$:

$$\frac{d\hat{T}}{dt} = -\frac{\overline{w}}{V\rho}\hat{T} + \frac{T_f - \overline{T}}{V\rho}\hat{w} + \frac{\hat{q}}{V\rho c_p} - \frac{2(\Delta H)k\overline{C}_A}{\rho c_p}\hat{C}_A \tag{9-9a}$$

$$\frac{d\hat{C}_A}{dt} = -\frac{\overline{w}}{V\rho}\hat{C}_A + \frac{C_{A_f} - \overline{C}_A}{V\rho}\hat{w} - 2k\overline{C}_A\hat{C}_A \tag{9-9b}$$

where
$$\hat{C}_A = C_A - \overline{C}_A$$
$$\hat{T} = T - \overline{T}$$
$$\hat{w} = w - \overline{w}$$
$$\hat{q} = q - \overline{q}$$

Equation 9-9 may be conveniently represented by the following matrix differential equation:

$$\mathbf{x}(t) = \mathbf{A}\mathbf{x}(t) + \mathbf{B}\mathbf{u}(t) \tag{9-10}$$

where

$$\mathbf{x}(t) = \begin{bmatrix} \hat{T} \\ \hat{C}_A \end{bmatrix} \quad \text{state vector}$$

$$\mathbf{u}(t) = \begin{bmatrix} \hat{q} \\ \hat{w} \end{bmatrix} = \text{manipulated inputs}$$

$$\mathbf{A} = \begin{bmatrix} -\dfrac{\overline{w}}{V\rho} & -\dfrac{2(\Delta H)k\overline{C}_A}{\rho c_p} \\ 0 & -\dfrac{\overline{w}}{V\rho} + 2k\overline{C}_A \end{bmatrix}$$

$$\mathbf{B} = \begin{bmatrix} \dfrac{1}{V\rho c_p} & \dfrac{T_f - \overline{T}}{V\rho} \\ 0 & \dfrac{C_{A_f} - \overline{C}_A}{V\rho} \end{bmatrix}$$

Equation 9-10 is simply a linear version of the state Eq. 9-2.

The measured outputs of a system may or may not be the state variables. In the above example, it is typically feasible to measure the reactor temperature (a state variable), and it could be an output. However, the reactor concentration is usually not easily measured. Instead, a measurable output could be a property such as viscosity or electrical conductivity which is a function of the concentration (and probably also the temperature). This relationship would become part of the function $\mathbf{g}$ in Eq. 9-4.

9-4 THE OPTIMAL CONTROL PROBLEM

In brief, the optimal control problem is to determine the control $\mathbf{u}(t)$ over the interval $t_0 < t \leqslant t_1$ that will transfer the system from an initial state $\mathbf{x}_0$ to some final state $\mathbf{x}(t_1)$ in such a manner as to minimize a performance functional. The final state $\mathbf{x}(t_1)$ may be explicitly specified, may be constrained in some manner, or may be completely unspecified. Similarly, the final time t_1 may or may not be specified. The performance functional J is in general an integral of a function L of the state $\mathbf{x}(t)$ and the control $\mathbf{u}(t)$, as represented by

$$J = \int_{t_0}^{t_1} L[\mathbf{x}(t),\mathbf{u}(t)]\, dt \tag{9-11}$$

More will be said about this later.

We shall consider two cases of the optimal control problem, one which shall be called the "fixed-end-point problem" and the other the "free-end-point problem." They are as follows (1).

Fixed-End-Point Problem

Let the system be described by the state equation

$$\dot{\mathbf{x}}(t) = \mathbf{f}[\mathbf{x}(t),\mathbf{u}(t)] \tag{9-12}$$

Let the initial state $\mathbf{x}(t_0)$ be specified at a specified time t_0. Furthermore, let the final state $\mathbf{x}(t_1)$ be specified, but let t_1 be free. Determine the control $\mathbf{u}(t)$ over the interval $t_0 < t \leqslant t_1$ that minimizes the cost functional J given by Eq. 9-11.

A special case of this problem is the minimum-time control problem, where it is desired to transfer the system from state $\mathbf{x}(t_0)$ to state $\mathbf{x}(t_1)$ in minimum time. In this case the cost functional becomes

$$J = \int_{t_0}^{t_1} L[\mathbf{x}(t),\mathbf{u}(t)]\, dt$$
$$= \int_{t_0}^{t_1} dt = t_1 - t_0 \tag{9-13}$$

Obviously the function L is simply unity.

Free-End-Point Problem

Let the system be described by the state equation

$$\dot{\mathbf{x}}(t) = \mathbf{f}[\mathbf{x}(t), \mathbf{u}(t)] \tag{9-14}$$

Let the initial state $\mathbf{x}(t_0)$ and initial time t_0 be specified. Determine the control $\mathbf{u}(t)$ over the interval $t_0 < t \leqslant t_1$, t_1 specified, that minimizes the cost functional J given by Eq. 9-11. Note that the final state $\mathbf{x}(t_1)$ is not specified.

The customary problem encountered under this case is the minimization of some quadratic performance criterion of the type

$$J = \int_{t_0}^{t_1} [\mathbf{x}^T(t)\mathbf{Q}\mathbf{x}(t) + \mathbf{u}^T(t)\mathbf{R}\mathbf{u}(t)]\, dt \tag{9-15}$$

where $\mathbf{Q}$ = positive definite matrix
$\mathbf{R}$ = positive semidefinite matrix

and superscript T denotes the transpose.

This performance functional is frequently used with linear systems because it gives relatively simple results. It is also possible for t_1 to be infinity.

9-5 THE MINIMUM PRINCIPLE

There seem to be at present three approaches to the solution of the optimal control problems formulated above:

1. Calculus of variations (5)
2. Pontryagin's minimum (or maximum) principle (6,7,8)
3. Dynamic programming (9,10)

The first of these encounters difficulties with such things as constraints on the manipulated variables, and does not appear to be of the practical utility of the last two. Thus it will not be considered. The minimum principle will be discussed in this section, with dynamic programming reserved for a subsequent section.

Before presenting the minimum principle, we must first define the Hamiltonian and the costate vector.

The Hamiltonian

Let $\mathbf{p}(t)$ be a vector whose order is the same as that of $\mathbf{x}(t)$. The Hamiltonian is a real-valued function defined as

$$H[\mathbf{x}(t), \mathbf{p}(t), \mathbf{u}(t)] = L[\mathbf{x}(t), \mathbf{u}(t)] + \mathbf{p}^T(t)\, \mathbf{f}[\mathbf{x}(t), \mathbf{u}(t)] \quad (9\text{-}16)$$

where $\mathbf{f}[\mathbf{x}(t), \mathbf{u}(t)]$ = right-hand side of the state equation
$L[\mathbf{x}(t), \mathbf{u}(t)]$ = integrand of the performance functional

Requiring that f and L be continuous and continuously differentiable with $\mathbf{x}$ means that the Hamiltonian will be likewise.

From the definition of the Hamiltonian and the state equation, it should be noted that

$$\dot{\mathbf{x}}(t) = \frac{\partial H[\mathbf{x}(t), \mathbf{p}(t), \mathbf{u}(t)]}{\partial \mathbf{p}(t)} = f[\mathbf{x}(t), \mathbf{u}(t)] \quad (9\text{-}17)$$

This equation together with the equation

$$\dot{\mathbf{p}}(t) = -\frac{\partial H[x(t), \mathbf{p}(t), \mathbf{u}(t)]}{\partial \mathbf{x}(t)} \quad (9\text{-}18)$$

are called the *canonical system of equations* for the optimal control problem. The vector $\mathbf{p}(t)$ is called the *costate vector* or the *adjoint*.

The minimum principle for our two problems is as follows:

Free End-Point Problem

The optimal control must be such that the following conditions are satisfied at the optimum.

1. The costate vector satisfies the canonical system of equations with the boundary conditions

$$\mathbf{x}(t_0) = \mathbf{x}_0$$
$$\mathbf{p}(t_1) = 0 \quad (9\text{-}19)$$
$$t_1 \text{ specified}$$

2. The Hamiltonian has an absolute minimum with respect to $\mathbf{u}(t)$.
3. The Hamiltonian is identically zero for the interval $t_0 < t \leqslant t_1$.

Fixed End-Point Problem

The optimal control must be such that the following conditions are satisfied at the optimum.

1. The costate vector satisfies the canonical system of equations with the boundary conditions

$$\mathbf{x}(t_0) = x_0 \qquad \mathbf{p}(t_1) = 0 \tag{9-20}$$

$$\mathbf{x}(t_1) = x_1 \qquad t_1 \text{ unspecified}$$

2. The Hamiltonian has an absolute minimum with respect to $\mathbf{u}(t)$.
3. The Hamiltonian is identically zero for the interval $t_0 < t \leqslant t_1$.

It should be noted that the only difference in the necessary conditions is the boundary conditions for the costate equations.

9-6 APPLICATION OF THE MINIMUM PRINCIPLE

The minimum principle is a very powerful and useful tool for determining the optimal control for problems falling into either of the above categories. Its application to process problems is beset by several difficulties. One of these lies with the cost functional. The natural cost functional to propose for process operation is to maximize the return or minimize the loss. However, the mathematical formulation of such a cost functional is not practical under most situations. The alternative generally selected is to substitute a cost functional that should give approximately the same results as one based on economics. The one frequently selected is minimum time.

To justify the reasoning behind this, suppose it is found that the process is currently operating at state $\mathbf{x}_0$. However, for current conditions, the optimal return would be for operation at state $\mathbf{x}_1$. Thus it seems reasonable to propose that the optimal control should transfer the process from $\mathbf{x}_0$ to $\mathbf{x}_1$ as soon as possible; i.e., in minimum time.

A second problem occurs in determining the optimal control from the minimum principle. While the minimum principle applies to nonlinear systems, to constraints on the manipulated variable, and other common complications encountered in process systems, constraints on the state variables, e.g., pressure or temperature limitations, cannot be readily incorporated. Even when these are absent, the computational requirements, especially for nonlinear systems, are considerable, basically due to the split boundary conditions on the canonical equation encountered in both cases considered in the last section.

A third difficulty arises from the fact that the minimum principle as formulated applies to what process engineers typically refer to as the open-loop control problem. That is, the minimum principle gives the control $\mathbf{u}$ as a function of time. For process systems, feedback control, i.e., control in which **u** is given as a function of the state **x**, is almost mandatory due to modeling errors, unknown disturbances, etc. Only for linear cases can a feedback control law be derived from the minimum principle with certainty.

Although these considerations reduce the utility of the minimum principle for process applications, it still offers a definite potential. Most of the above complications can be avoided if the minimum principle is applied to a simple, linear process model. Latour et al. (11) suggest that a model of wide utility is the following:

$$\frac{C(s)}{M(s)} = \frac{K_p e^{-\theta s}(\alpha s + 1)}{(\tau_1 s + 1)(\tau_2 s + 1)} \tag{9-21}$$

where τ_1, τ_2 = time constants
K = process gain
α = reciprocal of process zero
θ = dead time

Processes that can often be adequately represented by this model include extractors, mixing in agitated vessels, heat exchangers, distillation columns, and chemical reactors.

It thus seems reasonable to propose that an optimal control strategy for these units be based upon this model. The control problem is to drive the system from some known initial state $c(0)$, $\dot{c}(0)$ to some known final state $c(T)$, $\dot{c}(T)$ using a control subject to the constraint

$$U_{\min} \leqslant u \leqslant U_{\max}$$

The final time T is to be minimized. This is obviously identical to the "fixed-end-point problem" discussed above.

For the case in which both α and θ in Eq. 9-21 are zero, the control will always be at one of the extremes. For a second-order system, there will be two switches. If we let τ_1 be the larger of the time constants, the optimum switching times for the system are given by the equations in Table 9-1. Note that the equation for t_1 (the time at the first switch) requires an implicit solution. Figure 9-3 shows a typical input and a typical response. Note that t_1 and t_2 are the switching times; r_0 and r are the old and new set points, respectively; and K and k are the upper and lower constraints on the manipulated variable respectively. The response rises quickly but does not overshoot, which is typical of minimum time responses.

The application for which this procedure was proposed is for use in supervisory control. For example, suppose the computer calculates that for optimum operation the set point should be changed from r_0 to r. This transition should be made as follows:

TABLE 9-1
Switching Times for a Second-Order System
(For $r_0 < r$, interchange K and k in the equations below. These equations are specifically for the initial state $c(0) = r_0; \dot{c}(0) = 0$).

$r < r_0$
$0 < \tau_2/\tau_1 \leqslant 0.9$
$$\left[\frac{(K-k)-(r_0/K_p-k)\exp(-t_1/\tau_2)}{(K-r/K_p)}\right]^{\frac{\tau_2}{\tau_1}} = \frac{(K-k)-(r_0/K_p-k)\exp(-t_1/\tau_1)}{(K-r/K_p)}$$
$0.9 < \tau_2/\tau_1 \leqslant 1$
$$\left[\frac{r_0/K_p-k}{K-k}-\exp\left(\frac{t_1}{\tau_2}\right)\right]\ln\left[\frac{(r_0/K_p-k)-(K-k)\exp(t_1/\tau_2)}{r_0/K_p-K}\right]+\frac{t_1}{\tau_2}\exp\left(\frac{t_1}{\tau_2}\right)=0$$
$0 < \tau_2/\tau_1 \leqslant 1$
$$t_2/\tau_2 = \ln\left[\frac{(r_0/K_p-k)-(K-k)\exp(t_1/\tau_2)}{r/K_p-K}\right]$$

1. At time zero, the feedback controller should be placed on manual.
2. The manipulated variable should be switched from maximum to minimum or vice versa as discussed above.
3. At time t_2, the feedback controller should be returned to automatic.

Thus the feedback controller is present to "trim out" any modeling errors, load disturbances, and the like which may cause the optimal control to fall short of its stated objectives.

As for the case in which the process dead time is nonzero, consider the following representation of the process model:

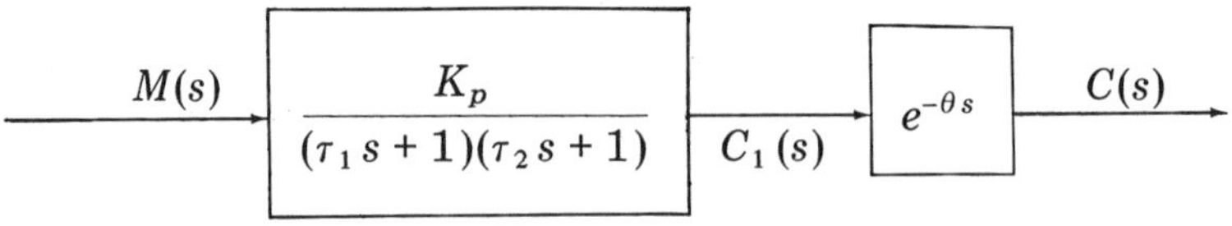

$$M(s) \longrightarrow \boxed{\frac{K_p}{(\tau_1 s+1)(\tau_2 s+1)}} \xrightarrow{C_1(s)} \boxed{e^{-\theta s}} \xrightarrow{C(s)}$$

Using the concepts presented above, the control $M(s)$ can be determined to give the optimum response $C_1(s)$ prior to the dead time. However, the dead time simply delays this response by time θ, which is completely independent of $M(s)$. Thus, the response $C(s)$ is op-

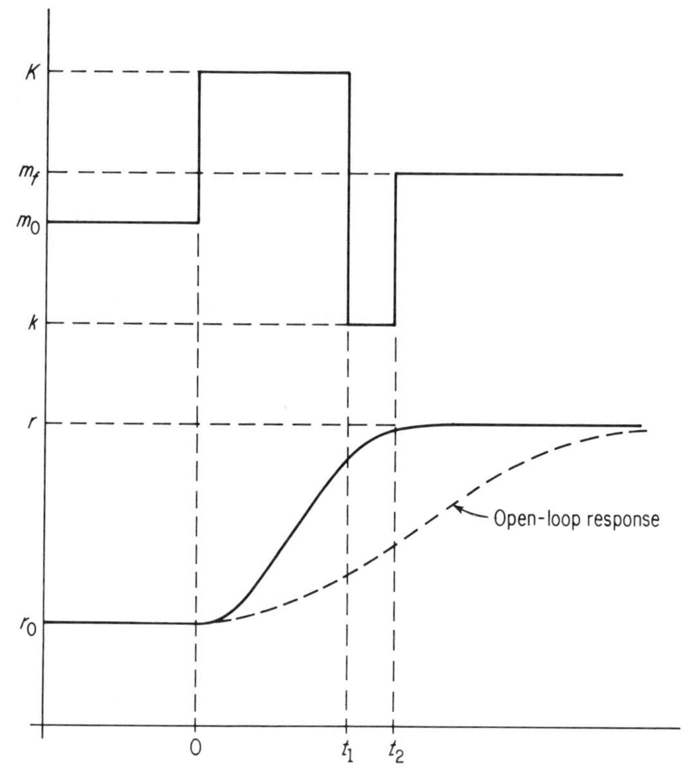

FIG. 9-3. Optimal response to a change in set point.

timized when $C_1(s)$ is optimized. In other words, the optimal control for the system with dead time is identical for the same system without the dead time. Note carefully that the above system is open loop, i.e., no feedback. The only modification to the control strategy in this paragraph is that the feedback controller should not be returned to automatic until time t_2 plus θ.

For cases in which α is not equal to zero, the optimum response is that the manipulated variable should follow a prescribed transient after the initial bang-bang action. As control of this type is difficult to achieve, Latour et al. (11) suggest the use of the same switching times presented above.

As pointed out previously, because of unmeasured load changes or other random disturbances, it is desirable to formulate the control strategy so that it can be implemented in a feedback manner. That is, we determine from the states $c(t)$ and $\dot{c}(t)$ if a switch should be made. This is readily implemented using a switching curve in the c-$\dot{c}$ plane as illustrated (for $\alpha = \theta = 0$) in Fig. 9-4. Note that the state

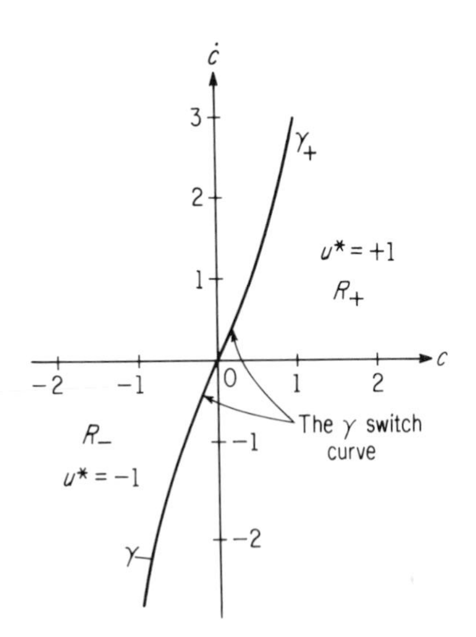

FIG. 9-4. Switching curve for a two-time-constant plant. (Reproduced by permission from M. Athans and P. L. Falb, *Optimal Control*, McGraw-Hill Book Company, New York, 1966.)

$c(t)$, $\dot{c}(t)$ specifies a location in the c-$\dot{c}$ plane. Depending upon the location of this point relative to the switching curve, the control u will be at one of its extremes. The procedure for developing the switching curve for the exact system considered is presented on pages 526–536 in Athan and Falb's book on optimal control (1).

Although a dead time θ in the process has no effect on the switching times, it will change the switching curve. As illustrated in Fig. 9-5, this is because the feedback is not the state vector $\mathbf{x}(t)$, but instead the delayed value $\mathbf{x}(t - \theta)$. Unfortunately, the method customarily used to determine the switching curve does not readily treat dead times in a direct fashion. Moore et al. (12) suggest incorporating the Smith predictor or dead time compensator (discussed in Sec. 8-8) as illustrated in Fig. 9-6a. Since this effectively moves the dead time

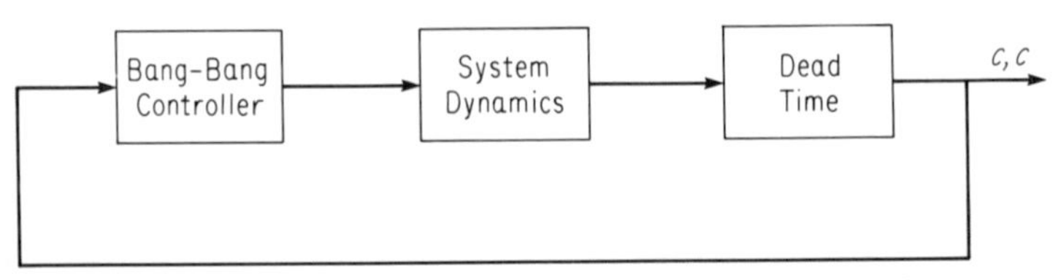

FIG. 9-5. Bang-bang control loop for systems with dead time.

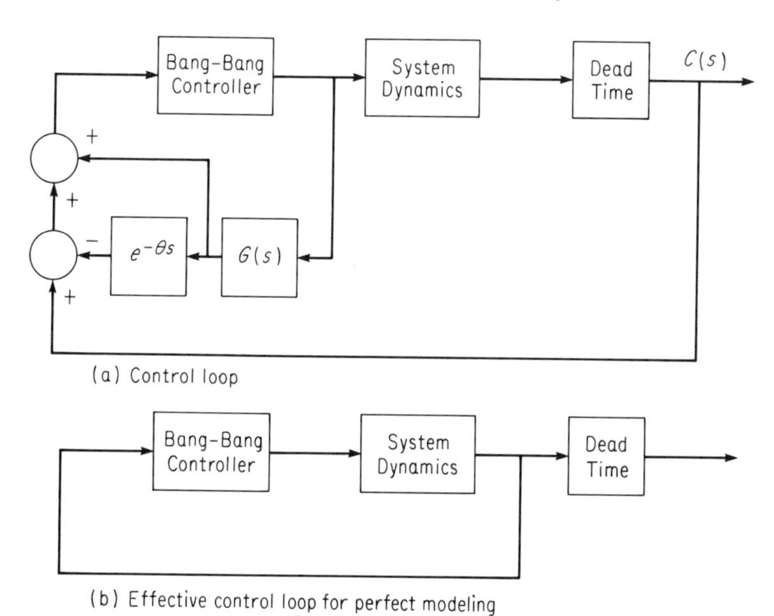

FIG. 9-6. Bang-bang controller coupled with the dead-time compensator.

outside the loop (Fig. 9-6b), the switching curve can be determined directly from the gains and time constants, ignoring the dead time. Note that the model required for the compensator is the same model used to determine the switching curve.

9-7 OPTIMAL CONTROL OF LINEAR SYSTEMS USING A QUADRATIC PERFORMANCE CRITERION (1)

This section will consider the optimal control of a linear, time invariant system given by the state equation

$$\dot{\mathbf{x}}(t) = \mathbf{A}\mathbf{x}(t) + \mathbf{B}\mathbf{u}(t) \tag{9-22}$$

$$\mathbf{x}(0) = \mathbf{x}_0 \tag{9-23}$$

It is desired to control the system in such a manner as to minimize the cost functional

$$J = \frac{1}{2}\int_0^T [\mathbf{x}^T(t)\mathbf{Q}\mathbf{x}(t) + \mathbf{u}^T(t)\mathbf{R}\mathbf{u}(t)]\,dt \tag{9-24}$$

This formulation is that of the state regulator problem, since in order to minimize the above cost functional the control will tend to drive

$\mathbf{x}(t)$ toward 0. The state equation further indicates that the equilibrium state corresponding to $\mathbf{x}(t)$ equal $\mathbf{0}$ is $\mathbf{u}(t)$ equal zero also.

To formulate the optimal control law, we begin by defining the Hamiltonian for this problem.

$$H = \frac{1}{2}\mathbf{x}^T(t)\mathbf{Q}\mathbf{x}(t) + \frac{1}{2}\mathbf{u}^T(t)\mathbf{R}\mathbf{u}(t) + \mathbf{p}^T(t)\mathbf{A}\mathbf{x}(t) + \mathbf{p}^T(t)\mathbf{B}\mathbf{u}(t) \tag{9-25}$$

The equation for the costate is

$$\dot{\mathbf{p}}(t) = \frac{\partial H}{\partial \mathbf{x}(t)} = -\mathbf{Q}\mathbf{x}(t) - \mathbf{A}^T\mathbf{p}(t) \tag{9-26}$$

From the minimum principle, the boundary condition should be

$$\mathbf{p}(t) = 0 \tag{9-27}$$

This equation and the state equation 9-22 form the canonical set of equations for this problem.

As presented in detail by Athans and Falb (1), the linearity of the canonical equations can be used to prove that the costate vector $\mathbf{p}(t)$ is a linear combination of the state vector $\mathbf{x}(t)$, or mathematically,

$$\mathbf{p}(t) = \mathbf{K}(t)\mathbf{x}(t) \tag{9-28}$$

This fact permits a reasonably simple solution to this optimal control problem.

In Sec. 9-3 it was noted that one of the requirements for $\mathbf{u}(t)$ to be optimal is that the Hamiltonian be minimized. Taking the partial of Eq. 9-25 and setting to zero gives

$$\frac{\partial H}{\partial \mathbf{u}(t)} = \mathbf{R}\mathbf{u}(t) + \mathbf{B}^T\mathbf{p}(t) = 0$$

or

$$\mathbf{u}(t) = -\mathbf{R}^{-1}\mathbf{B}^T\mathbf{p}(t) = -\mathbf{R}^{-1}\mathbf{B}^T\mathbf{K}(t)\mathbf{x}(t) \tag{9-29}$$

Thus we see that the control is also a linear function of the state, as illustrated by the feedback arrangement in Fig. 9-7.

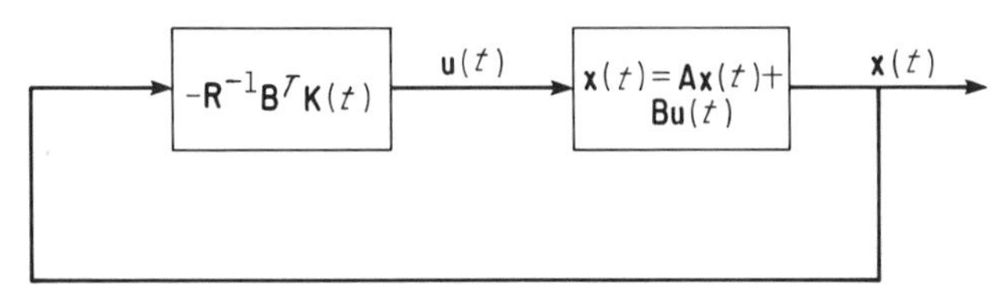

FIG. 9-7. Optimal controller.

Equation 9-29 is also quite suitable for a control law provided $\mathbf{K}(t)$ can be evaluated. To develop such a procedure, we begin by taking the derivative of Eq. 9-28:

$$\dot{\mathbf{p}}(t) = \mathbf{K}(t)\dot{\mathbf{x}}(t) + \dot{\mathbf{K}}(t)\mathbf{x}(t)$$

Substituting Eq. 9-26 for $\mathbf{p}(t)$ and Eq. 9-22 for $\mathbf{x}(t)$ followed by Eq. 9-28 for $\mathbf{p}(t)$ and Eq. 9-29 for $\mathbf{u}(t)$ gives ($\mathbf{K}(t)$ is also symmetric):

$$\dot{\mathbf{K}}(t) = -\mathbf{K}(t)\mathbf{A} - \mathbf{A}^T\mathbf{K}(t) + \mathbf{K}(t)\mathbf{B}\mathbf{R}^{-1}\mathbf{B}^T\mathbf{K}(t) - \mathbf{Q} \qquad (9\text{-}30)$$

This equation is known as the matrix Riccati equation, and can be solved for $\mathbf{K}(t)$ provided a boundary condition is available. From Eqs. 9-27 and 9-28 it is seen that

$$\mathbf{K}(T)\mathbf{x}(T) = 0$$

Since the final state $\mathbf{x}(T)$ is free (can assume any value), it follows that

$$\mathbf{K}(T) = 0 \qquad (9\text{-}31)$$

As this boundary condition is at the final time, Eq. 9-30 must be solved in reverse time to give $\mathbf{K}(t)$ over the interval $0 \leqslant t \leqslant T$.

A special case of interest is when $T \longrightarrow \infty$, or the control is over the infinite interval. For this case it can be shown that $\mathbf{K}(t)$ is a constant. Consequently, its derivative is zero, reducing Eq. 9-30 to

$$-\mathbf{K}\mathbf{A} - \mathbf{A}^T\mathbf{K} + \mathbf{K}\mathbf{B}\mathbf{R}^{-1}\mathbf{B}^T\mathbf{K} - \mathbf{Q} = 0 \qquad (9\text{-}32)$$

The only difficulty is in solving for $\mathbf{K}$. It turns out that a practical approach is to continue to use the differential equation 9-30 with the boundary condition of 9-31 and solve in reverse time until a "steady state" is reached, at which the value of $\mathbf{K}$ will be the solution to Eq. 9-32. This is illustrated in Fig. 9-8.

9-8 OPTIMAL CONTROL FOR SET-POINT CHANGES

The conventional control loop typically considered is illustrated in Fig. 9-9. The normal procedure is to design the controller either to a prescribed change in set point or to a prescribed change in disturbance (load). Unfortunately, the optimal controller as formulated in the previous section does not quite match either of these. Instead, it is designed to take the system from some initial state x_0 to the state 0 in an optimal fashion. In the remainder of this section and the next, we shall discuss the transformation of the conventional control problem into a form to which optimal control theory can be readily applied.

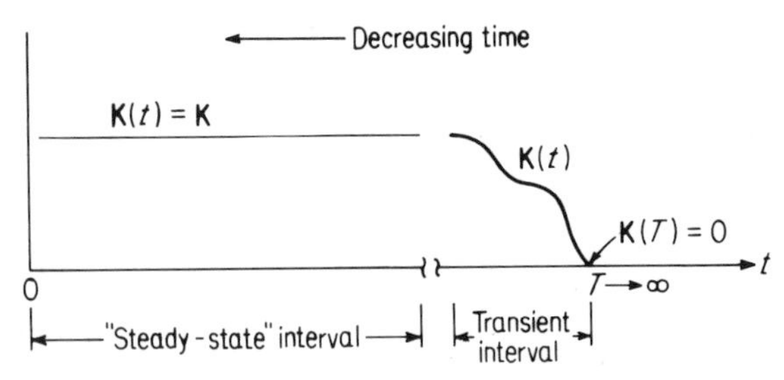

FIG. 9-8. A loose interpretation of the constant matrix K. As $T \longrightarrow \infty$, the "transient interval" tends to infinity and the "steady-state interval" occupies all finite times. (Reprinted by permission from M. Athans and P. L. Falb, *Optimal Control*, McGraw-Hill Book Company, New York, 1966.)

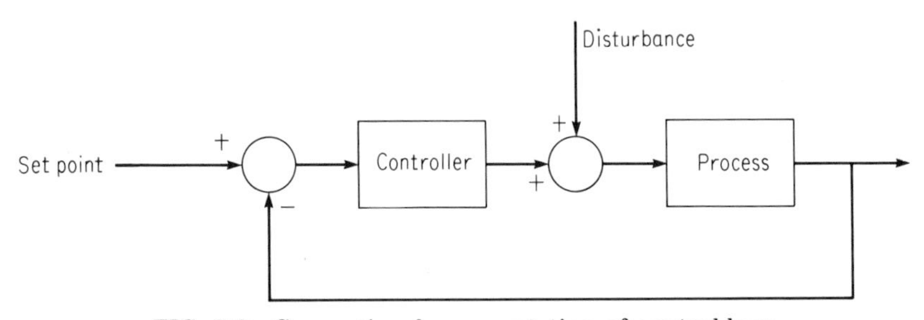

FIG. 9-9. Conventional representation of control loop.

We shall first consider the set point case. Specifically, suppose the first-order system

$$\frac{dx(t)}{dt} + x(t) = u(t) \tag{9-33}$$

is initially at state x_0. Suppose that at time zero the set point is changed to x_f. The typical response in this case is as shown in the top two graphs in Fig. 9-10.

To cast this problem into the optimal control formulation, it is necessary that the final value of the state variable be zero and the final value of the control be zero also. Thus we define two new variables as

$$x_1(t) = x(t) - x_f \tag{9-34}$$

$$u_1(t) = u(t) - u_f \tag{9-35}$$

Substituting into Eq. 9-33 gives

$$\frac{dx_1(t)}{dt} + x_1(t) + x_f = u_1(t) + u_f$$

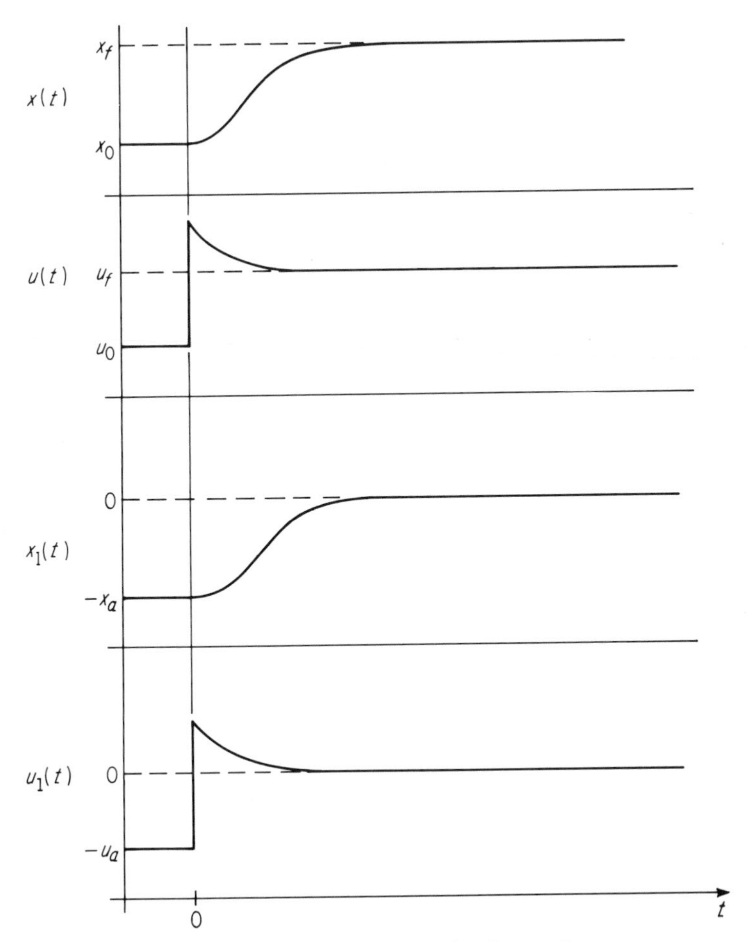

FIG. 9-10. Control and response for first-order system.

As Eq. 9-33 indicates that x_f equals u_f at steady state, this equation reduces to

$$\frac{dx_1(t)}{dt} + x_1(t) = u_1(t) \tag{9-36}$$

The boundary condition is

$$x_1(0) = x_0 - x_f = -x_a \tag{9-37}$$

As the control is now such that the state $x_1(t)$ is to be transferred from $-x_a$ (the initial condition) to zero (the origin), optimal control theory can be applied. Let the cost functional be defined as follows:

$$J = \frac{1}{2} \int_0^\infty [x_1(t)^2 + u_1(t)^2]\, dt \tag{9-38}$$

Substituting for the corresponding quantities in Eq. 9-32 gives

$$+ 2k + k^2 - 1 = 0$$

The solution is

$$k = 0.416$$

Thus the controller is a pure proportional controller with a gain of 0.416.

Here we begin to have some difficulties. Using a pure proportional controller, we are proposing to make a set point change and not have any offset (i.e., error) at the new operating point. The only case in which the proportional control will not exhibit such offset is at its equilibrium point. By making the above change of variable, we effectively defined this equilibrium point to be at the new set point.

It is also interesting to note that the controller does not exhibit the integral mode. As the control is simply a linear combination of the states of the system (see Eq. 9-29), we will have an integral mode only if we define a state corresponding to the integral of the state variable. For the first-order system considered above, this could potentially be accomplished by the approach in Fig. 9-11. The

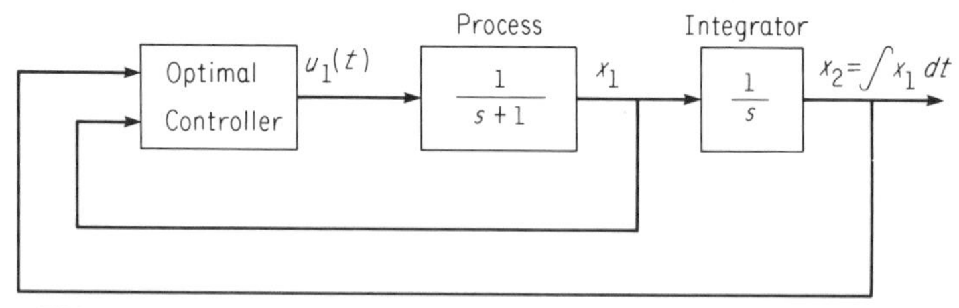

FIG. 9-11. A possible means for introducing an integral term into the optimal control law.

performance functional must be of the form

$$J = \int_0^{\infty} [q_1 x_1^2(t) + q_2 x_2^2(t) + u^2(t)]\, dt \tag{9-39}$$

The difficulty arises in assigning a "cost" to the state $x_2(t)$ which corresponds to the integral mode, i.e., select a value for q_2. Since this mode was added with the supposition that it could be used in the control law to achieve better control and is not part of the original process, it seems reasonable to set q_2 equal to zero. However, this leads to a zero value of the gain corresponding to the integral state variable, thus defeating the purpose for which the integral state was originally proposed.

9-9 OPTIMAL CONTROL TO DISTURBANCE CHANGES

As the example to illustrate how an optimal controller may be designed for disturbance changes (13), consider the system in Fig. 9-12a. The state equation describing the process for this case is

$$\dot{x}_1(t) = -x_1(t) + d(t) + u(t) \tag{9-40}$$

$$x_1(0) = 0 \tag{9-41}$$

Note that the disturbance appears as an input along with the control $u(t)$. To be cast into the optimal control formulation, we must transform the problem in such a manner that the disturbance appears as an initial condition.

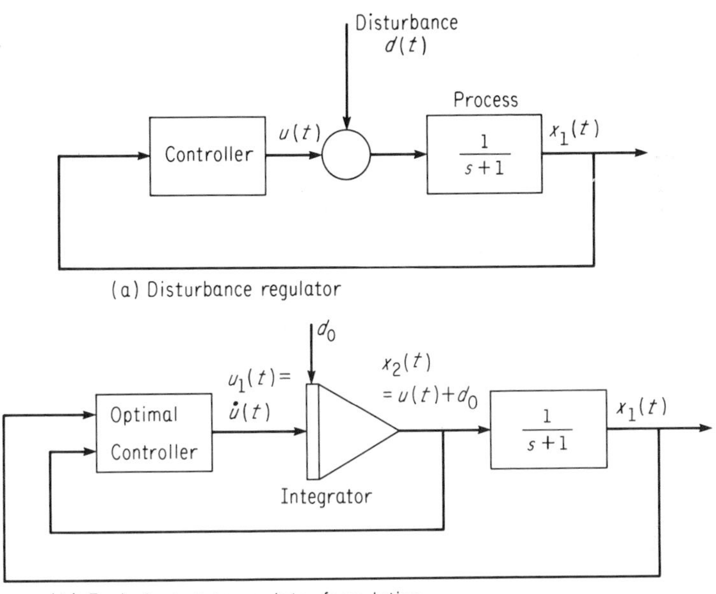

FIG. 9-12. Transformation of the conventional disturbance regulator control problem into the optimal state regulator problem.

For the specific case in which the disturbance is a step change, this may be accomplished by the formulation in Fig. 9-12b. If the disturbance is a step change from 0 to d_0 at time zero, this may effectively appear as an initial condition on an integrator. If the continuous input to the integrator is $\dot{u}(t) = u_1(t)$, the output is the sum of $d(t)$ and $u(t)$, as illustrated in Fig. 9-12b.

From this point, we proceed as usual. First, note that the state equations are (in terms of the new variables).

$$\dot{x}_1(t) = -x_1(t) + x_2(t) \tag{9-42}$$

$$\dot{x}_2(t) = u_1(t) \tag{9-43}$$

$$x_1(0) = 0$$

$$x_2(0) = d_0$$

In matrix form, this becomes

$$\frac{d}{dt}\begin{bmatrix} x_1(t) \\ x_2(t) \end{bmatrix} = \begin{bmatrix} -1 & 1 \\ 0 & 0 \end{bmatrix}\begin{bmatrix} x_1(t) \\ x_2(t) \end{bmatrix} + \begin{bmatrix} 0 \\ 1 \end{bmatrix} u_1(t) \tag{9-44}$$

The cost functional could be

$$J = \int_0^\infty [x_1^2(t) + u_1^2(t)]\, dt \tag{9-45}$$

Although the proper matrices could be substituted into Eq. 9-32, the resulting equations cannot be analytically solved for the coefficients of matrix **K**. Instead, the solution of Eq. 9-30 in backward time until steady-state is reached will yield the solution

$$\mathbf{K} = \begin{bmatrix} k_{11} & k_{12} \\ k_{21} & k_{22} \end{bmatrix}$$

Substituting into Eq. 9-29 gives

$$u_1(t) = -k_{21}x_1(t) - k_{22}x_2(t) \tag{9-46}$$

Again the control is a linear function of the states.

However, in this case the integrator is not really part of the process, but a part of the controller instead. Therefore we may eliminate $x_2(t)$ by substituting Eq. 9-42 into Eq. 9-46. Also noting that $u_1(t)$ is really $\dot{u}(t)$ gives

$$\dot{u}(t) = -k_{21}x_1(t) - k_{22}[x_1(t) + \dot{x}_1(t)]$$

Integrating gives

$$u(t) = -k_{22}x_1(t) - (k_{21} - k_{22}) \int_0^t x_1(\tau)\, d\tau + U_0 \tag{9-47}$$

where U_0 = constant of integration. Thus we have proportional-plus-integral control.

It should also be noted that the cost functional in Eq. 9-45 is actually

$$J = \int_0^\infty [x_1^2(t) + \dot{u}^2(t)]\, dt$$

That is, the cost functional penalizes changes in control rather than for actual magnitude.

The application of the approach to control one of the unit operations has been reported by Miller (14). The system was a simulated distillation column, subjected to feed disturbances. The boilup rate was ratioed to the feed rate, and a feedback controller regulated the distillate rate to control the overheads composition (15). The scheme is illustrated in Fig. 9-13.

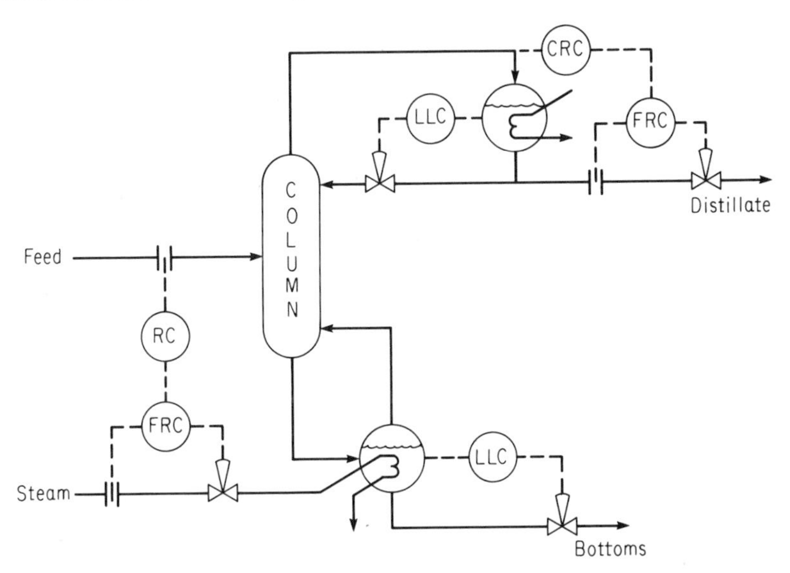

FIG. 9-13. Control scheme for distillation column.

As the manipulated variable is the distillate rate, a transfer function is needed to relate changes in overhead composition to changes in distillate rate. In Laplace transform notation, this model is

$$Y_D(s) = G_1(s)D(s) + F_e(s)$$

where $F_e(s)$ is the effect of a given change in feed rate on the overhead composition. From step responses such as those given in Fig. 9-14, it is apparent that $G_1(s)$ is a first-order lag for all practical purposes. Basing the time constant on the 63.2 percent point and the gain on the final steady-state values, averaging the values for the four responses in Fig. 9-14 gives the following model:

$$Y_D(s) = \frac{-0.0140}{0.55s + 1} D(s) + F_e(s)$$

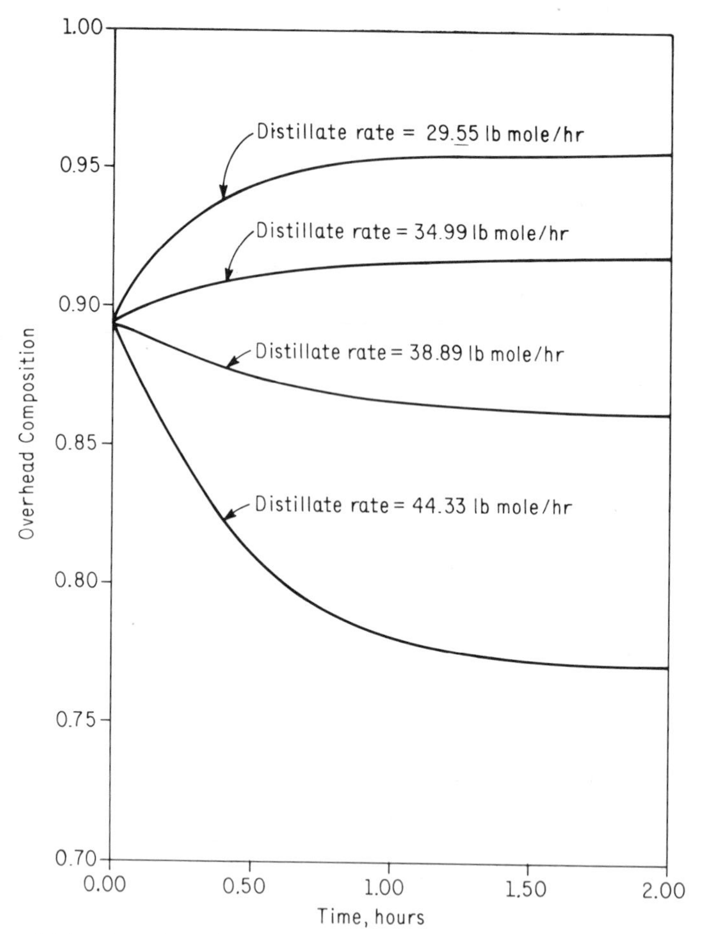

FIG. 9-14. Open-loop responses to several step changes in distillate rate.

Expressing in state-variable form gives the equation

$$\dot{y}_D(t) = \frac{dy_D(t)}{dt} = \frac{-1}{0.55}\, y_D(t) + \frac{-0.0140}{0.55}\, D(t) + \frac{1}{0.55}\, F_e(t)$$

Using the approach outlined previously in this section, the criterion function should be

$$J(D) = \int_0^\infty \{[y_{D_{set}} - y_D(t)]^2 + r\dot{D}^2\}\, dt$$

The cost $J(D)$ consists of two parts. The first part $[y_{D_{set}} - y_D(t)]^2$ penalizes for deviations of the controlled variable y_D from its desired value $y_{D_{set}}$. The second part D^2 penalizes for changes in the manipulated variable D.

Applying the method presented previously gives the following control law.

$$D(t) = K_2 \int_0^t [y_{D_{set}} - y_D(\tau)]\, d\tau + K_1[y_{D_{set}} - y_D(t)] + D_0$$

where K_1, K_2 = control parameters
D_0 = constant of integration

The optimal controller is the familiar proportional plus integral feedback controller.

Figure 9-15 shows the effect of r on the resulting responses of y_D, the controlled variable, and D, the manipulated variable, to a step change in the feed rate. As would be suspected, small values of r lead to tight control and large changes in D. Large values of r produce the opposite results. Thus the parameter r is essentially a tuning parameter whose value must be determined by experimenting with the process in much the same manner as current controllers are tuned.

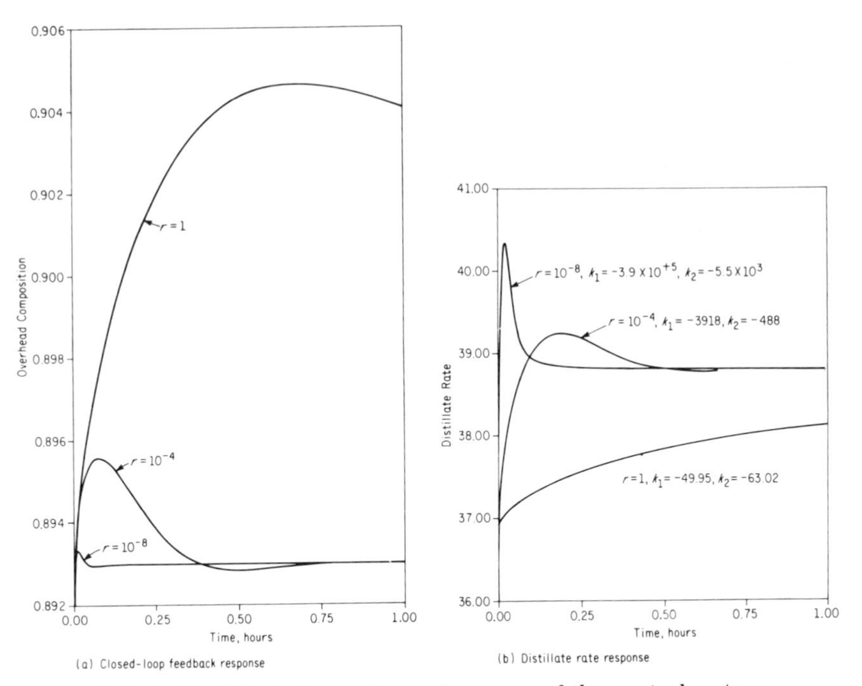

FIG. 9-15. Effect of r on the performance of the control system.

Before concluding, it should be pointed out that the control strategy is optimal for disturbance changes, but is not necessarily optimal for set-point changes. Furthermore, use of higher-order process models will produce higher derivative terms in the control strategy (i.e., a second-order model will yield a PID controller). Processes with dead times could be a potential problem, but perhaps dead-time compensation (16) can be used to surmount this difficulty.

9-10 DYNAMIC PROGRAMMING

The optimization procedure which most nearly meets the feedback requirements outlined in Sec. 9-1 is dynamic programming (9, 10). Bellman (9) states the underlying concepts in his principle of optimality, which states that whatever the initial state and initial decisions are, the optimal policy is such that all remaining decisions must constitute an optimal policy with regard to the state resulting from the first decision. This is very similar to the feedback principle of basing all control decisions on the state of the system.

Dynamic programming is most readily formulated for problems with decisions at discrete points (stages) such as the transportation problem considered in Sec. 9-1. Sampled data or digital control systems also fall nicely into this category, as decisions must be made only at the sampling times. True continuous problems must be discretized so that they too fall into this category.

Suppose the system begins at some stage 0 and passes through a succession of n stages to reach the final destination. The primary relationship used in the dynamic programming formulation simply states that the cost over the remainder of the control path consists of the sum of the cost of traversing from stage i to stage $i + 1$ plus the cost of traversing from stage $i + 1$ to stage n, the final stage. To illustrate the logic further, consider the following representation of the problem.

Suppose the system is at state x_i at the ith stage in time. Suppose the control u_i is applied over the next sampling instant. Then the state x_{i+1} is related to x_i and u_i by the discrete state equation

$$x_{i+1} = f(x_i, u_i) \tag{9-48}$$

Suppose the cost incurred over this sampling time as a consequence of this control is $h(x_i, u_i)$. Furthermore, let the minimum cost of the traverse between state x_{i+1} at the $(i + 1)$st sampling time and the final destination be given by $g_{i+1}(x_{i+1})$. That is, no matter how we reach state x_{i+1}, we shall use the optimal strategy from there to the final state.

We shall now collect all of these terms into a relationship that gives the cost $g_i(x_i)$ of traversing from the ith sampling time to the final destination. The total cost of traversing from i to n is the sum of the cost from i to $i + 1$ plus the cost from $i + 1$ to n, or mathematically,

$$\text{Cost from } i \text{ to } n = h(x_i, u_i) + g_{i+1}(x_{i+1}) \tag{9-49}$$

As the state x_{i+1} is expressed in terms of the state x_i and the control u_i by the discrete state equation 9-48, x_{i+1} can be eliminated from Eq. 9-49 to give

$$\text{Cost from } i \text{ to } n = h(x_i, u_i) + g_{i+1}[f(x_i, u_i)] \tag{9-50}$$

The cost $g_i(x_i)$ of optimal trajectory from state x_i, no matter how we got there, to the terminal state is given by

$$g_i(x_i) = \min_{u_i} \{h(x_i, u_i) + g_{i+1}[f(x_i, u_i)]\} \tag{9-51}$$

For the state x_i there is thus a corresponding control u_i (depending only upon the state x_i) that gives the minimum cost $g_i(x_i)$ over the remainder of the trajectory. Equation 9-51 is simply a recurrence relationship giving the cost $g_i(x_i)$ in terms of $g_{i+1}(x_{i+1})$. This recurrence relationship is initialized by

$$g_n(x_n) = \text{specified} \tag{9-52}$$

That is, the cost at the terminal point is known.

To illustrate these points, suppose we consider determining the optimal control strategy for the first-order system

$$\tau \frac{dx(t)}{dt} + x(t) = u(t) \tag{9-53}$$

with the initial state 1.0 at time zero. Let the sampling time be 1.0, the control u being held constant between sampling times. The optimal trajectory should be such that the cost functional

$$J = x_3^2 + \sum_{i=0}^{2} (x_i^2 + u_i^2) \tag{9-54}$$

is minimized. This cost functional differs from previous examples in that there is a terminal cost, that is, a cost associated with the final state.

We begin by first determining the discrete state equation corresponding to Eq. 9-53. Using the techniques outlined in Sec. 7-3, this equation is

$$x_{i+1} = (1 - e^{-T/\tau})x_i + (e^{-T/\tau})u_i$$

Letting τ equal 0.608 for numerical convenience ($e^{-T/\tau} = 0.5$), this equation becomes

$$x_{i+1} = 0.5x_i + 0.5u_i \tag{9-55}$$

From Eq. 9-54, the function $h(x_i, u_i)$ is given by

$$h(x_i, u_i) = x_i^2 + u_i^2$$

and final cost

$$g_3(x_3) = x_3^2$$

We can now proceed with the application of the recurrence relationship in Eq. 9-51.

At the sampling instant $i = 2$, the cost is given by

$$h(x_2, u_2) + g_3(x_3) = x_2^2 + u_2^2 + x_3^2$$

Substituting the discrete state equation for x_3 and minimizing with respect to u_2 gives

$$g_2(x_2) = \min_{u_2} \{x_2^2 + u_2^2 + (0.5x_2 + 0.5u_2)^2\}$$

This minimum can be evaluated by taking the derivative with respect to u_2 and setting equal to zero, the result being

$$u_2 = -0.2x_2 \tag{9-56}$$

This is the control law by which the control decision is to be made at the sampling instant $i = 2$. The cost associated with this decision is

$$g_2(x_2) = 1.2x_2^2$$

This we have established $g_2(x_2)$ and u_2 in terms of only the state x_2.

Using these same procedures, the equations for the remaining sampling instants are

$i = 1$:

$$h(x_1, u_1) + g_2(x_2) = x_1^2 + u_1^2 + 1.2x_2^2$$

$$g_1(x_1) = \min_{u_1} \{x_1^2 + u_1^2 + 1.2(0.5x_1 + 0.5u_1)^2\}$$

$$u_1 = -0.23x_1 \tag{9-57}$$

$$g_1(x_1) = 1.231x_1^2$$

$i = 0$:

$$h(x_0, u_0) + g_1(x_1) = x_0^2 + u_0^2 + 1.231x_1^2$$

$$g_0(x_0) = \min_{u_0} \{x_0^2 + u_0^2 + 1.231(0.5x_0 + 0.5u_0)^2\}$$

$$u_0 = -0.236x_0 \tag{9-58}$$

$$g_0(x_0) = 1.236x_0^2$$

As the initial state is $x_0 = 1$, the minimum cost for the trajectory is 1.236. Equations 9-56, 9-57, and 9-58 are the control laws to be used at each sampling instant. These should clearly illustrate that dynamic programming generates the feedback-control law directly, i.e., control equals a function of the state.

The above example was purposely chosen to have an analytic solution. Of course, most practical problems would not fall into this category. Instead of having analytic functions at each stage, the cost $g_i(x_i)$ and the control u_i would be a tabulated function of x_i, i.e., of the following form as illustrated for $i = 1$ for the above problem:

x_1	$g_1(x_1)$	u_1
0.0	0.	0.0
0.2	0.2462	-0.0462
0.4	0.4924	-0.0924
0.6	0.7386	-0.1386
0.8	0.9848	-0.1848
1.0	1.2310	-0.2310

By using linear interpolation, this table can serve just as well for determining u_1 from the state x_1 as will the analytic function 9-57 in a practical case.

The cost $g_1(x_1)$ being of tabular nature also eliminates the possibility of minimizing the cost $g_0(x_0)$ in the next stage by taking the analytic derivative of Eq. 9-51. This minimization must now be undertaken by a numerical search procedure. This is, however, a one-dimensional search that can be readily executed.

In general, dynamic programming is attractive because it is basically a numerical procedure that is readily programmed on a digital computer, much more so than the solution of the two-point boundary value problem arising from the minimum principle.

If dynamic programming is so attractive, why has it not been more widely used in the process control field? The answer lies in the dimensionality of the problems. The above problem had only one state variable, which was discretized to give six discrete values for storage in the computer. Six discrete values are probably not enough, 20 to 50 being more reasonable. However, for simplicity, suppose we take ten discrete values. If we have four state variables in a problem, this means that we must store $10^4 = 10{,}000$ values of g_i and 10,000 values of u_i at each stage. If 100 discrete values were stored, this number would be 10 million. Bellman has called this the "curse of dimensionality." Although some progress has been made to reduce this difficulty, it still limits the cases to which dynamic programming can be successfully applied.

9-11 IN SUMMARY

Hopefully this chapter has provided some insight into how the somewhat unique requirements of process control are reflected in the applications of optimal control theory. Another new development on the horizon is estimation theory (17). Although only a few applications have appeared (18, 19, 20), certainly more will appear in the future.

LITERATURE CITED

1. Athans, M., and P. L. Falb, *Optimal Control*, McGraw-Hill, New York, 1966.
2. Lee, E. B., and L. Markus, *Foundations of Optimal Control Theory*, Wiley, New York, 1967.
3. Sage, A. P., *Optimum Systems Control*, Prentice-Hall, Englewood Cliffs, N.J., 1968.
4. Koppel, L. B., *Introduction to Control Theory*, Prentice-Hall, Englewood Cliffs, N.J., 1968.
5. Hestenes, M. R., *Calculus of Variations and Optimal Control Theory*, Wiley, New York, 1966.
6. Pontryagin, L. S., et al., *The Mathematical Theory of Optimal Processes*, Interscience, New York, 1962.
7. Rozonoer, L. T., "The Mathematical Theory of L. S. Pontryagin in Optimal-Systems Theory," *Automat. Telemech.*, Moscow, Vol. 20, (1969), pp. 1320, 1441, 1561.
8. Fan, L. T., *The Continuous Maximum Principle*, Wiley, New York, 1966.
9. Bellman, R. E., *Dynamic Programming*, Princeton U. P., Princeton, N.J., 1957.
10. Roberts, S. M., *Dynamic Programming in Chemical Engineering and Process Control*, Academic Press, New York, 1964.
11. Latour, P. R., L. B. Koppel, and D. R. Coughanowr, "Time Optimum Control of Chemical Processes for Set-Point Changes," *I & EC Process Design and Development*, Vol. 6, No. 4 (October 1967), p. 452.
12. Moore, C. F., C. L. Smith, and P. W. Murrill, "Control of a High Order Plant Using a Time Optimal Second-Order Switching Curve," *ISA Trans.*, Vol. 8, No. 3 (1969), pp. 186–191.
13. Johnson, C. D., "Optimal Control of the Linear Regulator with Constant Disturbances," *IEEE Trans. on Automatic Control*, August 1968, p. 416.
14. Miller, J. A., C. L. Smith, and P. W. Murrill, "An Application of Optimal Control Theory," presented at the 25th Annual ISA Conference and Exhibit, Philadelphia, October 1970.
15. Shinskey, F. G., *Process Control Systems*, McGraw-Hill, New York, 1967, p. 288.
16. Smith, O. J. M., "Close Control of Loops with Dead Time," *Chemical Engineering Progress*, Vol. 53, No. 5 (May 1967), pp. 217–219.

17. Meditch, J. S., *Stochastic Optimal Linear Estimation and Control*, McGraw-Hill, New York, 1969.
18. Wells, C. H., and L. E. Larson, "Combined Optimum Control and Estimation of Serial Systems with Time Delay, presented at the 1969 JACC, Boulder, Colorado, Aug. 5-7, 1969.
19. Wells, C. H., "Application of Modern Estimation and Identification Techniques to Chemical Process," presented at the 1969 JACC, Boulder, Colorado, Aug. 5-7, 1969.
20. ——, "Application of Estimation Theory to the BOF," *Proceedings* of the Fifth Annual Workshop on the Use of Digital Computers for Process Control, LSU, Baton Rouge, Feb. 4-6, 1970.

appendix **A**

Table of z-Transforms

Table of z-Transforms

Time Function, $f(t)$	Laplace Transform, $F(s)$	z-Transform, $F(z)$	Modified z-Transform, $F(z, m)$
$u(t)$	$\frac{1}{s}$	$\frac{1}{1 - z^{-1}}$	$\frac{z^{-1}}{1 - z^{-1}}$
t	$\frac{1}{s^2}$	$\frac{Tz^{-1}}{(1 - z^{-1})^2}$	$\frac{mTz^{-1}}{1 - z^{-1}} + \frac{Tz^{-2}}{(1 - z^{-1})^2}$
t^2	$\frac{2!}{s^3}$	$\frac{T^2 z^{-1}(1 + z^{-1})}{(1 - z^{-1})^3}$	$T^2\left[\frac{m^2 z^{-1}}{1 - z^{-1}} + \frac{(2m + 1)z^{-2}}{(1 - z^{-1})^2} + \frac{2z^{-3}}{(1 - z^{-1})^3}\right]$
t^{n-1}	$\frac{(n - 1)!}{s^n}$	$\lim_{a \to 0} (-1)^{n-1} \frac{\partial^{n-1}}{\partial a^{n-1}}\left(\frac{1}{1 - e^{-aT}z^{-1}}\right)$	$\lim_{a \to 0} (-1)^{n-1} \frac{\partial^{n-1}}{\partial a^{n-1}}\left(\frac{e^{-amT}z^{-1}}{1 - e^{-aT}z^{-1}}\right)$
e^{-at}	$\frac{1}{s + a}$	$\frac{1}{1 - e^{-aT}z^{-1}}$	$\frac{e^{-amT}z^{-1}}{1 - e^{-aT}z^{-1}}$
$\frac{1}{b - a}(e^{-at} - e^{-bt})$	$\frac{1}{(s + a)(s + b)}$	$\frac{1}{b - a}\left(\frac{1}{1 - e^{-aT}z^{-1}} - \frac{1}{1 - e^{-bT}z^{-1}}\right)$	$\frac{z^{-1}}{b - a}\left(\frac{e^{-amT}}{1 - e^{-aT}z^{-1}} - \frac{e^{-bmT}}{1 - e^{-bT}z^{-1}}\right)$
$\frac{1}{a}(u(t) - a^{-aT})$	$\frac{1}{s(s + a)}$	$\frac{1}{a}\frac{(1 - e^{-aT})z^{-1}}{(1 - z^{-1})(1 - e^{-aT}z^{-1})}$	$\frac{z^{-1}}{a}\left(\frac{1}{1 - z^{-1}} - \frac{e^{-amT}}{1 - e^{-aT}z^{-1}}\right)$
$\frac{1}{a}\left(t - \frac{1 - e^{-at}}{a}\right)$	$\frac{1}{s^2(s + a)}$	$\frac{1}{a}\left[\frac{Tz^{-1}}{(1 - z^{-1})^2} - \frac{(1 - e^{-aT})z^{-1}}{a(1 - z^{-1})(1 - e^{-aT}z^{-1})}\right]$	$\frac{z^{-1}}{a}\left[\frac{T}{(1 - z^{-1})^2} + \frac{amT - 1}{a(1 - z^{-1})} + \frac{e^{-amT}}{a(1 - e^{-aT}z^{-1})}\right]$

$\frac{(a-b)}{a^2}u(t)+\frac{b}{a}t+\frac{1}{a}\left(\frac{b}{a}-1\right)e^{-at}$	$\frac{s+b}{s^2(s+a)}$	$\frac{z^{-1}}{a}\left[\frac{bT}{(1-z^{-1})^2}+\frac{(a-b)(1-e^{-aT})}{a(1-z^{-1})(1-e^{-aT}z^{-1})}\right]$	$\frac{z^{-1}}{a}\left[\frac{bTz^{-1}}{(1-z^{-1})^2}+\left(bmT+1-\frac{b}{a}\right)\frac{1}{1-z^{-1}}+\frac{b-a}{a}\frac{e^{-amT}}{1-e^{-aT}z^{-1}}\right]$
$\frac{1}{ab}\left(u(t)+\frac{b}{a-b}e^{-at}-\frac{a}{a-b}e^{-bt}\right)$	$\frac{1}{s(s+a)(s+b)}$	$\frac{1}{ab}\left[\frac{1}{1-z^{-1}}+\frac{b}{(a-b)(1-e^{-aT}z^{-1})}-\frac{a}{(a-b)(1-e^{-bT}z^{-1})}\right]$	$\frac{z^{-1}}{ab}\left[\frac{1}{1-z^{-1}}+\frac{be^{-amT}}{(a-b)(1-e^{-aT}z^{-1})}-\frac{ae^{-bmT}}{(a-b)(1-e^{-bT}z^{-1})}\right]$
te^{-at}	$\frac{1}{(s+a)^2}$	$\frac{Te^{-aT}z^{-1}}{(1-e^{-aT}z^{-1})^2}$	$\frac{Te^{-amT}z^{-1}\left[m+(1-m)e^{-aT}z^{-1}\right]}{(1-e^{-aT}z^{-1})^2}$
$\sin at$	$\frac{a}{s^2+a^2}$	$\frac{z^{-1}\sin aT}{1-2z^{-1}\cos aT+z^{-2}}$	$\frac{z^{-1}\sin amT+z^{-2}\sin(1-m)aT}{1-2z^{-1}\cos aT+z^{-2}}$
$\frac{1}{b}e^{-at}\sin bt$	$\frac{1}{(s+a)^2+b^2}$	$\frac{1}{b}\frac{z^{-1}e^{-aT}\sin bT}{1-2z^{-1}e^{-aT}\cos bT+e^{-2aT}z^{-2}}$	$\frac{z^{-1}e^{-amT}}{b}\frac{\sin bmT+z^{-1}e^{-aT}\sin(1-m)bT}{1-2z^{-1}e^{-aT}\cos bT+e^{-2aT}z^{-2}}$
$e^{-at}\cos bt$	$\frac{s+a}{(s+a)^2+b^2}$	$\frac{1-z^{-1}e^{-aT}\cos bT}{1-2z^{-1}e^{-aT}\cos bT+e^{-2aT}z^{-2}}$	$\frac{e^{-amT}z^{-1}\left[\cos bmT+z^{-1}e^{-aT}\sin(1-m)bT\right]}{1-2z^{-1}e^{-aT}\cos bT+e^{-2aT}z^{-2}}$
$\cos at$	$\frac{s}{s^2+a^2}$	$\frac{1-z^{-1}\cos aT}{1-2z^{-1}\cos aT+z^{-2}}$	$\frac{z^{-1}\cos amT-z^{-2}\cos(1-m)aT}{1-2z^{-1}\cos aT+z^{-2}}$

For a more extensive table see B. C. Kuo, *Analysis and Synthesis of Sampled-Data Control Systems*, Prentice-Hall, Inc. Englewood Cliffs, N.J., 1963.)

appendix B

Exercises

CHAPTER 4

4-1. Using the definition of the z-transform, verify that $\mathfrak{z}\{t\} = \dfrac{Tz}{(z-1)^2}$.

4-2. Using the definition of the z-transform, show that

$$\mathfrak{z}\{f(t+T)\} = zF(z) - zf(0)$$

4-3. Prove the initial-value theorem

$$f(0) = \lim_{z \to \infty} F(z)$$

4-4. The final-value theorem for sampled-data systems is

$$f(\infty) = \lim_{t \to \infty} f(t) = \lim_{z \to 1} \{(1 - z^{-1})\, F(z)\}$$

provided $F(z)$ contains no poles on or outside the unit circle. To prove this theorem, first show that

$$\lim_{z \to 1} [\mathfrak{z}\{f(t+T) - f(t)\}] = f(\infty) - f(0)$$

and then take the indicated transforms on the left-hand side of the equation.

4-5. Suppose the process pulse transfer function $HG(z)$ is represented as follows:

$$HG(z) = \frac{P(z)}{Q(z)} = \frac{p_0 + p_1 z^{-1} + \cdots + p^m z^{-m}}{q_0 + q_1 z^{-1} + \cdots + q^n z^{-n}}$$

Show that the gain K of the process is given by the expression

$$K = \frac{\sum_{i=0}^{m} p_i}{\sum_{i=0}^{m-k} q'_i}$$

where k = type of the system and

$$Q'(z) = \frac{Q(z)}{(1 - z^{-1})^k} = q'_0 + q'_1 z^{-1} + \cdots + q'_{n-k} z^{-(n-k)}$$

4-6. Suppose a controller $D(z)$ is represented by the following transfer function:

$$D(z) = \frac{K(1 + \alpha_1 z^{-1})(1 + \alpha_2 z^{-1})}{(1 + a_1 z^{-1})(1 + a_2 z^{-1})(1 + a_3 z^{-1})}$$

Show that in order to remove a pole or zero without changing the gain of the controller, it is only necessary to let z equal one in the term to be removed. That is, to remove the pole at $z = -a_1$ and not change the gain, the new control algorithm should be:

$$D(z) = \frac{K}{1 + a_1} \frac{(1 + \alpha_1 z^{-1})(1 + \alpha_2 z^{-1})}{(1 + a_2 z^{-1})(1 + a_3 z^{-1})}$$

4-7. What is the ultimate gain for the control system considered in Sec. 4-6? That is, what is the largest value of gain that will give stable behavior?

4-8. Rework the example in Sec. 4-6 with $R(s) = 0$ and $N(s) = 1/s$. Determine the output for the first five sampling instants for $T = 1$ and $K = 0.5$. Is there a steady-state offset [i.e., does $c(\infty) = r(\infty)$]?

4-9. Rework the example in Sec. 4-6 except use the PI control law and the process transfer function $G(s) = \frac{1}{s+1}$. Determine the difference equation relating the output c to the input r.

4-10. In Sec. 4-8 the transfer function for the block diagram in Fig. 4-8 was derived to be

$$\frac{C(z)}{R(z)} = \frac{z^{-1}(1.125 + 0.455 z^{-1})}{1 + 0.757z^{-1} + 0.455z^{-2}}$$

Plot the pole-zero locations. Is the system stable?

4-11. Verify that

$$\mathfrak{z}_m\{t\} = \frac{mT}{z-1} + \frac{T}{(z-1)^2}$$

4-12. Verify that

$$\mathfrak{z}\{f(t)\} = z\mathfrak{z}_m\{f(t)\}|_{m=0}$$

4-13. For the control loop in Fig. 4-9, the closed-loop transfer function was derived to be:

$$\frac{C(z)}{R(z)} = \frac{z^{-1}(1.125 + 0.455z^{-1})}{1 + 0.757z^{-1} + 0.455z^{-2}}$$

Determine $c(nT)$, $n = 0, 1, \ldots, 5$, if $R(z)$ is the unit step function.

4-14. Use the modified z-transform to determine the inter-sample behavior for $c(t)$ in the block diagram in Fig. 4-8, where

$$D(z) = K, K = 0.5$$
$$H(s) = (1 - e^{-sT})/s$$
$$G(s) = 1/s$$
$$T = 1.0$$
$$R(s) = 1/s$$
$$N(s) = 0$$

4-15. Derive the pulse transfer function for a process-zero-order hold combination where the process transfer function is given by:

$$G(s) = \frac{e^{-0.78s}}{4s^2 + 3.6s + 1}$$

Let the sampling time equal 0.5.

CHAPTER 5

5-1. Suppose a signal is generated as follows:

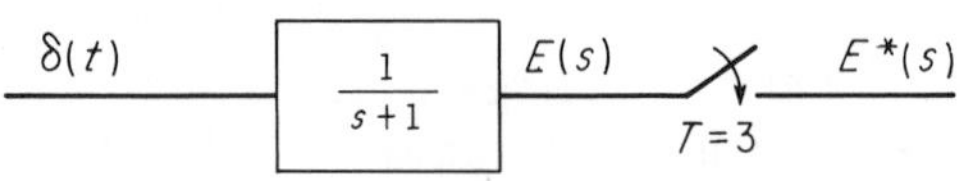

$\delta(t)$ → $\frac{1}{s+1}$ → $E(s)$ → sampler ($T = 3$) → $E^*(s)$

PROB. 5-1

Plot $E(j\omega)$ and $E^*(j\omega)$.

5-2. The fractional-order hold is represented as follows:

$$m(t) = m(nT) + k\,\frac{m(nT) - m[(n-1)T]}{T}\,(t - nT)$$

Its output is shown in the accompanying diagram.

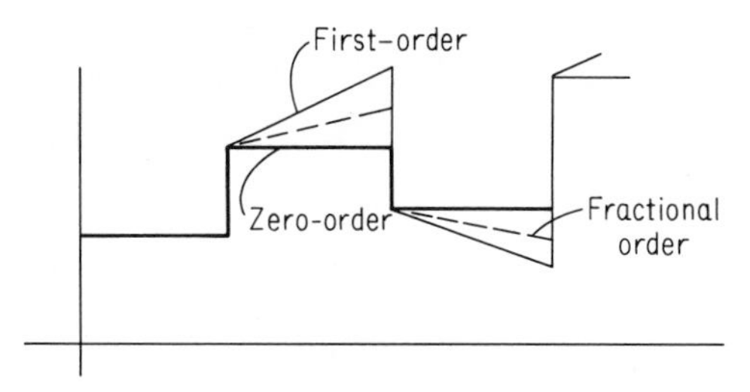

PROB. 5-2

Derive the transfer function for this hold.

5-3. Rework the example in Sec. 4-6 except use a first-order hold instead of the zero-order hold.

5-4. Rework the example in Sec. 4-6 except use a continuous slewer in place of the zero-order hold.

5-5. Derive the transfer function for the block diagram in Fig. 4-8 if a first-order hold were used.

5-6. The damping ratio is a commonly used design criterion for continuous systems. For sampled data systems, suppose we require that no poles in the primary strip have a damping ratio ζ less than, say, 0.5. That is, all poles must be enclosed by the following contour:

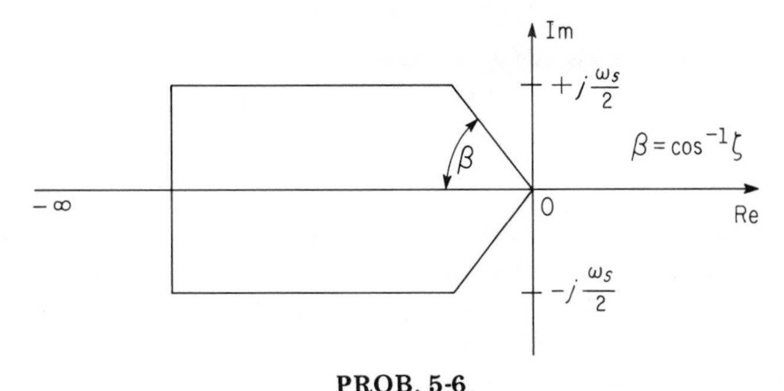

PROB. 5-6

Plot this contour in the z-domain.

5-7. Consider the block diagram shown in the accompanying illustration. Determine the pulse transfer function $HG(z)$ and compute the ultimate gain via stability considerations in the z-domain. Alternatively use the dead-time approximation to the sample and hold and compute the ultimate gain for the resulting control loop (frequency response concepts readily permit this). Do the two results agree?

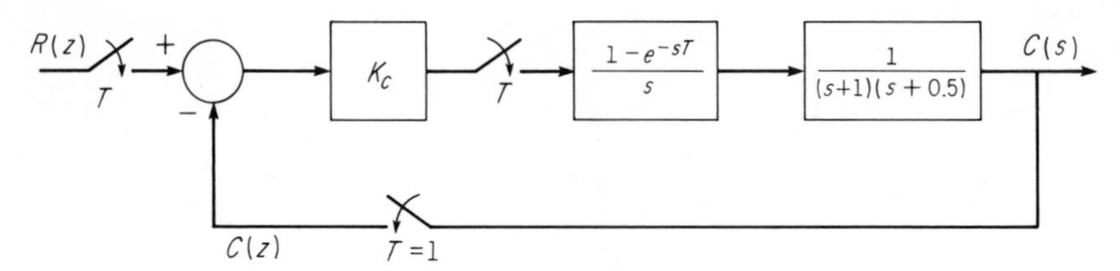

PROB. 5-7

5-8. Let the controller in Fig. 4-8 be a pure proportional instrument with gain K_c. Thus the closed-loop block diagram becomes

$$\frac{C(z)}{R(z)} = \frac{K_c HG(z)}{1 + K_c HG(z)}$$

where

$$HG(z) = \frac{z^{-1}(0.450 + 0.182z^{-1})}{1 - 0.368z^{-1}}$$

A root locus diagram can be constructed for $HG(z)$ in exactly the same manner as one is constructed for a continuous transfer function. Draw the root locus and determine the ultimate gain for the closed-loop system.

5-9. Recall that the Routh-Hurwitz criterion could be used to determine if any poles of a continuous system are in the right-half plane. For sampled-data systems, the transformation

$$z = \frac{r + 1}{r - 1}$$

maps the inside of the unit circle into the left-half r-plane. Determine the ultimate gain for the system of Exercise 5-8 by making this transformation and applying the Routh-Hurwitz criterion to the result.

5-10. Alternatively, the transformation

$$z = \frac{1 + w}{1 - w}$$

can be used to accomplish the same objective as the transformation in the previous exercise. Determine the ultimate gain by this approach.

5-11. To develop the discrete-filtering equation using z-transforms, the system to be examined is the following:

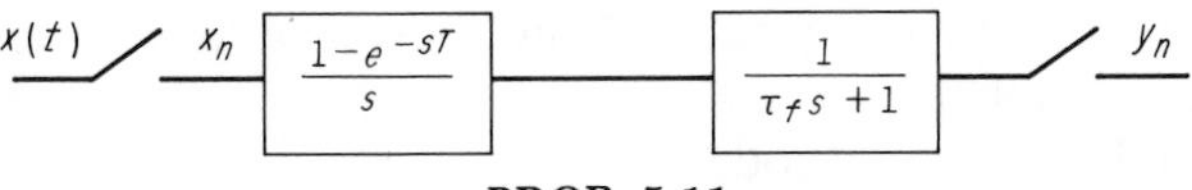

PROB. 5-11

The filtering equation is the difference equation corresponding to the pulse transfer function

$$\mathfrak{z}\left[\frac{1-e^{-sT}}{s}\cdot\frac{1}{\tau_f s+1}\right]$$

CHAPTER 6

Instead of individual exercises for this chapter, more insight can be obtained from a case study. For this the students will need a digital computer at their disposal and a multivariable search routine. For one or more of the three systems described below, one or more of the following aspects could be assigned to the students working individually or in groups:

1. Model the system by each applicable shortcut technique described in Sec. 6-1. For each method, plot the model response and the process response. Compute the sum of squares of the model error.
2. Using a nonlinear regression package or multivariable search technique, determine the best possible fit for each model considered in Part 1. Plot the model response and the process response. Compare the sum of squares to that obtained in Part 1.
3. Again using the nonlinear regression package or multivariable search technique, determine the best possible fit for a model whose order is one greater than that of the highest order model considered in Part 2. Again plot and compare results. Increase the order again if it appears that a significant reduction in the sum of squares could be obtained.
4. In each of the above three parts, the model was fit to the step response. Repeat Part 2 but using the response from a square pulse, i.e., the input $u(t)$ to be used is

$$u(t)=\begin{cases}0 & t<0, t\geqslant t_c\\ 1 & 0\leqslant t<t_c\end{cases}$$

   Compare the resulting models to those determined in Part 2.
5. Using the better models determined in the above steps, de-

sign controllers by each of the following techniques:
(a) Deadbeat
(b) Dahlin's method
(c) Kalman's method
Eliminate ringing if it is a problem.

6. For cases in which the manipulated variable would saturate, the velocity form of the PI or PID algorithm was recommended. What should be done in the case of saturation for algorithms considered in Part 5? Plot the results for Dahlin's method in Part 5.
7. For each of the algorithms considered in Part 5, determine the response to a step change in the load input.
8. For the tuning methods Z-N, C-C, IAE, and ITAE (both load and set point), determine the response to both changes in load and changes in set point. For each method, compute the IAE, the ITAE, and the decay ratio for both PI and PID control.
9. Determine the "optimum" settings based on a second-order lag plus dead time model using IAE and ITAE. Consider both PI and PID control and both load and set-point inputs. Compare to the results obtained in Part 8.
10. Determine the true optimum settings for the process being considered. Again, consider both PI and PID control and both load and set-point inputs. Compare to the results obtained in Parts 7 and 8.
11. For a step change in set point, determine the settings that give no overshoot and minimum rise time. Consider both PI and PID control. First, determine the settings based on a first-order model; then a second-order model; and finally, the process itself.

For the above investigations, one of the following systems could be simulated:

System A: Jacketed Chemical Reactor

In the jacketed chemical reactor described in the accompanying illustration, the second-order exothermic reaction $2A \rightarrow B$ proceeds. For purposes of deriving the equations, the reactor and the jacket are assumed to be well mixed. The dynamics of the sensor are described by a time constant τ_t and the dynamics of the flow control loop are represented by a time constant τ_v. The system is subject to abrupt changes of 5° in set point or 200 lb/min in the reactant flow rate.

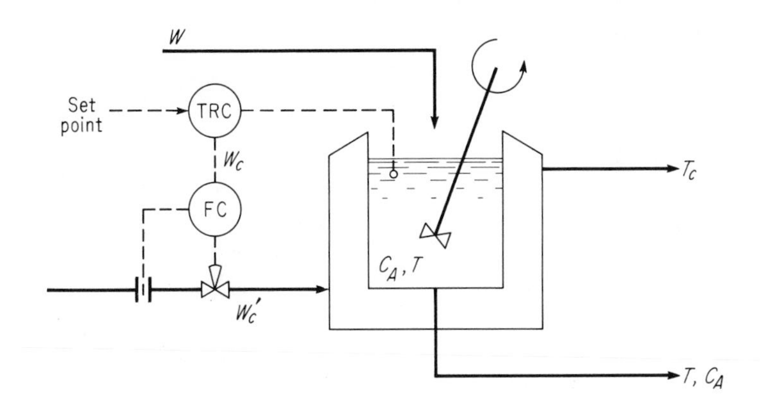

Definition of Variables

a	Constant in Arrhenius expression for reaction rate constant, 2560°R
A	Effective jacket heat transfer area, 500 sq ft
C_A	Concentration of reactant A in reactor and exit stream, lb/cu ft
C_{Af}	Concentration of A in feed, lb/cu ft
c_p	Specific heat or reacting mixture, 0.9 Btu/lb °F
c_p'	Specific heat of water, 1.0 Btu/lb °F
$-\Delta H$	Heat or reaction, 867 Btu/lb of A
k	Reaction rate constant, ft^3/lb-min
k_0	Constant in Arrhenius expression, 1.43 ft^3/lb-min
M_c	Mass of the jacket water, 4,000 lb
t	Time, min
T	Reactor temperature, °F
T'	Measured temperature of the reactor, °F
T_c	Jacket temperature, °F
T_f	Feed temperature, °F
U	Heat transfer coefficient, 1.2 Btu/min-sq ft-°F
V	Reactor volume, 250 cu ft
W	Feed rate, lb/min
W_c	Cooling water rate, as set by control algorithm, lb/min
W_c'	Cooling water rate, after control-loop lag, lb/min
ρ	Density of reacting mixture, 80 lb/cu ft
τ_t	Time constant of temperature transmitter, 1.0 min
τ_v	Time constant of flow control loop, 0.1 min

Steady-State Conditions

T = 190°F
k = 0.0278 ft^3/lb-min

$W = 1{,}000$ lb/min
$T_f = 150°$ F
$T_c = 120°$ F
$C_{Af} = 9.0$ lb/ft³
$C_A = 3.6$ lb/ft³
$W_c = 1{,}050$ lb/min
$T_w = 80°$ F

Equations

Mass balance on A

$$\frac{dC_A}{dt} = \frac{W}{V\rho}(C_{A_f} - C_A) - kC_A^2$$

Enthalpy balance on reacting mass

$$V\rho c_p \frac{dT}{dt} = Wc_p(T_f - T) - UA(T - T_c) + (-\Delta H)VkC_A^2$$

Enthalpy balance on jacket

$$M_c c_p' \frac{dT_c}{dt} = UA(T - T_c) - W_c' c_p' (T_c - T_w)$$

Dependence of rate constant on temperature

$$k = k_0 \exp\left[-a/(T + 460)\right]$$

Dynamics of temperature sensor

$$\tau_t \frac{dT'}{dt} + T' = T$$

Dynamics of flow controller

$$\tau_v \frac{dW_c'}{dt} + W_c' = W_c$$

System B: Vaporizer

The system shown in the accompanying sketch supplies a vapor at a constant rate V mole/hr to another unit in the process. The feed is a hydrocarbon mixture at temperature T_f. To keep the system reasonably simple, we will assume it contains only two components, propane and butane, at mole fractions x_{1_f} and x_{2_f}, respectively. To prevent buildup of minor components, material is purged at rate P mole/hr. Steam of enthalpy H is used as the heating medium. The design of a control algorithm for the PRC is the item of interest.

The following assumptions are made for this system:

1. The flow controller and the level controller act instantaneously; i.e., their dynamics can be neglected.

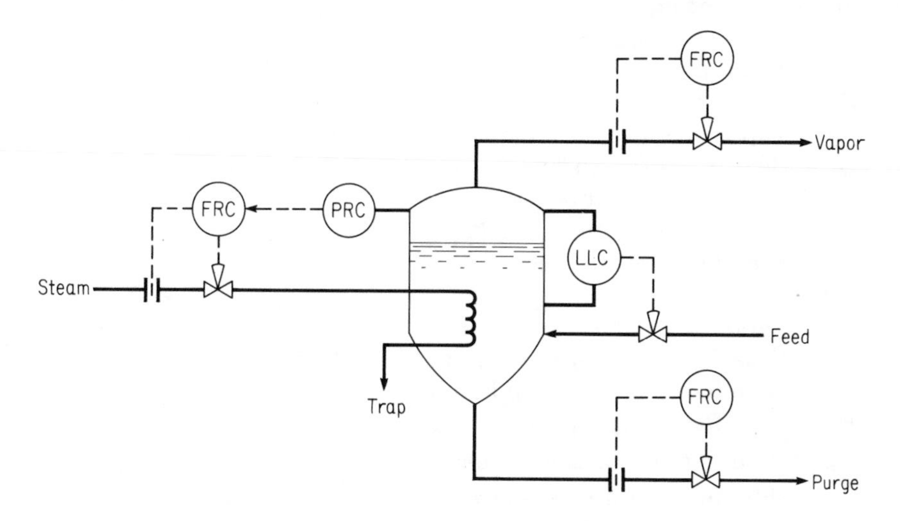

2. Holdup in the vapor phase is negligible.
3. All physical properties (density, heat capacity, etc.) are independent of composition and temperature.
4. The liquid is assumed to be ideal, and thus to follow Raoult's law. Vapor pressures will be calculated from the Antoine equation. Although vapor-liquid equilibrium can be computed more accurately by other means, this approach is adequate for our purposes and is quite simple to use.
5. The steam chest is assumed to have a time constant τ_c.
6. Only the latent heat of vaporization of the steam is transferred to the liquid.

This system is subject to three significant input changes:

1. Changes in the set point of the PRC.
2. Changes in the vapor rate V.
3. Changes in the feed compositions.

Abrupt changes of 10 percent in each are not unusual.

Definition of Variables

A_1 constant in Antoine equation, 6.82973
A_2 constant in Antoine equation, 6.83029
B_1 constant in Antoine equation, 813.20
B_2 constant in Antoine equation, 882.80

C_1	constant in Antoine equation, 248.00
C_2	constant in Antoine equation, 240.00
c_p	Average molar heat capacity, 29.0 Btu/mole °F
F	Hydrocarbon feed rate, mole/hr
F_s	Steam rate, lb/hr
M	Moles of liquid in vessel, 100.0 moles
P	Purge rate, 1.0 mole/hr
P_T	Total pressure, psia
P_1	Vapor pressure of propane, psia
P_2	Vapor pressure of butane, psia
Q	Heat transfer, Btu/hr
T	Liquid temperature, °F
T_c	Liquid temperature, °C; $T_c = (T - 32)/1.8$
T_f	Feed temperature, °F
t	Time, hours
V	Vapor rate, mole/hr
x_1	Mole fraction propane in liquid
x_2	Mole fraction butane in liquid
x_{1_f}	Mole fraction propane in feed
x_{2_f}	Mole fraction butane in feed
y_1	Mole fraction propane in vapor
y_2	Mole fraction butane in vapor
λ_1	Latent heat of vaporization of propane, 4750 Btu/mole
λ_2	Latent heat of vaporization of butane, 7890 Btu/mole
λ_s	Latent heat of vaporization of steam, 952 Btu/lb
τ_c	Time constant of steam chest, 1.0 min

Steady-State Conditions

F	1,000 mole/hr
F_s	8,146 lb/hr
P_T	318 psia
T	162.2°F
T_f	80°F
V	999 mole/hr
x_1	0.673
x_2	0.327
x_{1_f}	0.8
x_{2_f}	0.2

Derivation of Equations

Total material balance

$$F = V + P$$

Component material balance (component 1)

$$Fx_{1_f} - Px_1 - Vy_1 = M\frac{dx_1}{dt}$$

Enthalpy balance

$$Fc_p(T_f - T) - Vy_1\lambda_1 - Vy_2\lambda_2 = Mc_p\frac{dT}{dt}$$

Vapor chest

$$\tau_c\frac{dQ}{dt} + Q = \lambda F_s$$

Sum of mole fractions

$$x_2 = 1 - x_1$$

Antoine equations

$$\log_{10}\left(\frac{760P_1}{14.7}\right) = A_1 - B_1/(C_1 + T_c)$$

$$\log_{10}\left(\frac{760P_2}{14.7}\right) = A_2 + B_2/(C_2 + T_c)$$

Total pressure

$$P_T = P_1x_1 + P_2x_2$$

Vapor composition

$$y_1 = P_1x_1/P_T$$

$$y_2 = 1 - y_1$$

System C: Heater

A heat-sensitive liquid stream is exposed for a brief period of time to a set of steam-heated coils in the apparatus illustrated in

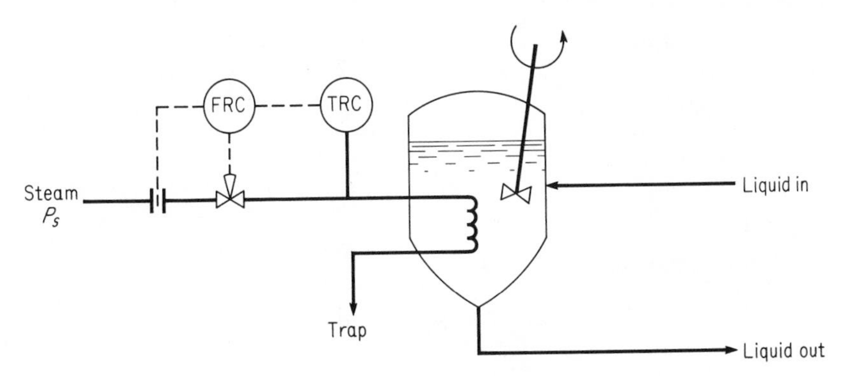

the accompanying figure. In order to prevent thermal decomposition on the coils, the temperature inside the coils is carefully regulated. The system is subjected to sudden increases or decreases in liquid flow rate of up to 20 percent.

Assumptions

1. Enthalpy of steam in the coils is negligible compared to enthalpy of the coils.
2. The quantity of liquid in the vessel is constant and is well mixed.
3. Only the latent heat of vaporization of the steam is recovered.

Definition of Variables

A	Heat transfer area, 100 ft^2
c_p	Liquid heat capacity, 0.6 Btu/lb °F
c_p'	Heat capacity of metal in coil, 0.2 Btu/lb °F
F	Liquid feed rate, lb/hr
F_s	Steam rate, lb/hr
F_s'	Set point on flow controller, lb/hr
M	Mass of the coils, 100 lb
M_L	Mass of liquid in vessel, 500 lb
Q	Heat transfer rate, Btu/hr
T	Steam temperature, °F
T'	Measured steam temperature, °F
T_f	Liquid feed temperature, 60° F
T_L	Liquid temperature, °F
T_R	Reference temperature, 32° F
t	Time, min
λ_s	Latent heat of vaporization of the steam, 870 Btu/lb
τ_b	Time constant of the bulb measuring the steam temperature, 0.2 min
τ_F	Time constant of the flow control loop, 0.5 min

Steady State Conditions:

F	10,000 lb/hr
F_s	673.5 lb/hr
Q	600,000 Btu/hr
T	220° F
T_L	160° F

Equations

Enthalpy balance on coils

$$F_s\lambda_s - Q = Mc_p' \frac{dT}{dt}$$

Heat transfer

$$Q = UA(T - T_L)$$

Enthalpy balance on liquid

$$Fc_p(T_f - T_L) + Q = M_L c_p \frac{dT_L}{dt}$$

Dynamics of flow control loop

$$\tau_F \frac{dF_s}{dt} + F_s = F'$$

Dynamics of measurement bulb

$$\tau_b \frac{dT'}{dt} + T' = T$$

CHAPTER 7

7-1. Use the modified z-transform to develop the linear regression formulation for a first-order lag plus dead time when the dead time is not restricted to multiples of the sampling time.

7-2. Develop the linear regression formulation for a second-order lag, first-order lead system.

7-3. Equations 7-6, 7-12, and 7-14 were presented as alternate forms of the linear regression formulation. Would you expect to obtain the same parameter values from each formulation?

7-4. When formulating the regression equations via z-transforms, we have been using the zero-order hold, which gives the following type of reconstruction (solid line):

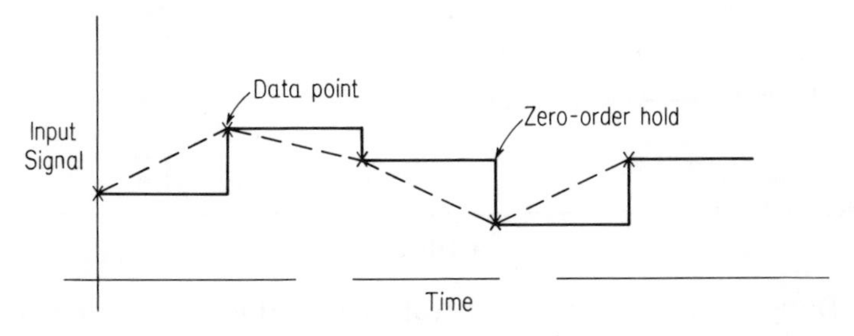

PROB. 7-4

In many cases, a better reconstruction could be obtained if a straight line were drawn between the data points, as illustrated by the dashed line. The Laplace transform expression for this type of hold is

$$H(s) = \frac{e^{sT}}{T}\left(\frac{1 - e^{-sT}}{s}\right)^2$$

Formulate the linear difference equation for a first-order process $\frac{K}{\tau s + 1}$ using this type of reconstruction. Can linear regression be used to determine values of K and τ from input/output data?

CHAPTER 8

8-1. One very common industrial application of feedforward control is in pH control systems. In the paper "Adaptive Feedback Applied to Feedforward pH Control" by Shinskey and Myron (Proceedings of the 25th Annual ISA Conference, Philadelphia, October 1970, Paper 565), a nonlinear feedforward control system is suggested. Develop the block diagram for the entire process and the recommended control system.

8-2. A pure hydrogen recycle stream is to be combined with a hydrocarbon vapor stream (containing some hydrogen) to produce a stream containing 90 percent by volume of hydrogen. Determine if feedforward control can be used to improve the performance of the scheme shown in the accompanying figure if (a) flow rate of hydrocarbons is constant but its composition (percent hydrogen) varies; (b) composition of the hydrocarbon stream is constant but the flow rate varies; (c) both composition and flow rate vary.

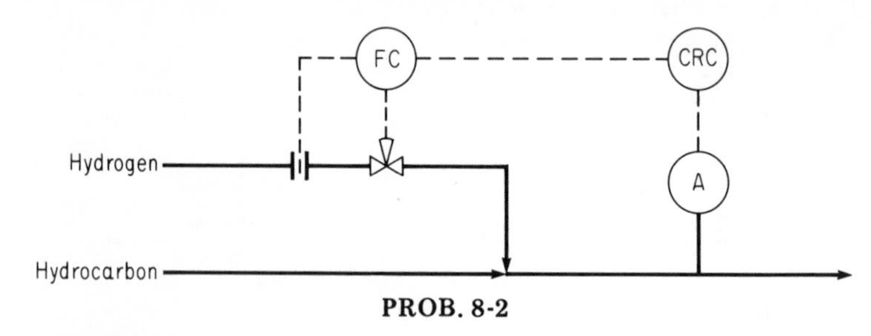

PROB. 8-2

8-3. Design a feedforward control system that would improve the performance of the control system for the heater (System C) considered in the exercises for Chapter 6.

8-4. Incorporate a feedforward strategy into the control system in Fig. 8-9 to compensate for changes in feed flow rate.

8-5. The simulation of System A for Chapter 6 could readily be used to ascertain how much improvement can be obtained from cascade control as compared to straight feedback control for this specific system.

8-6. Suppose we have the following process:

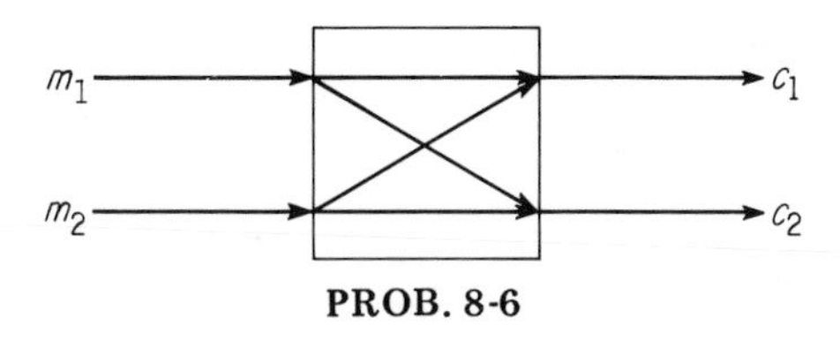

PROB. 8-6

A unit step increase in m_1 gives the following steady-state changes in the outputs:

c_1	6.1
c_2	1.7

A unit step change in m_2 gives the following steady-state changes in the output:

c_1	-1.2
c_2	0.8

Suggest the proper pairing of controlled and manipulated variables. Design a decoupler in which the gains on the diagonal (with proper pairing) are unity.

8-7. Catalyst A with a variable amount of carrier C is mixed with reactant B, and fed to the flash tank illustrated in the figure where the carrier C is vaporized (some B is carried over). It is desired to control the total liquid flow out of the tank as well as the concentrations of A and B.

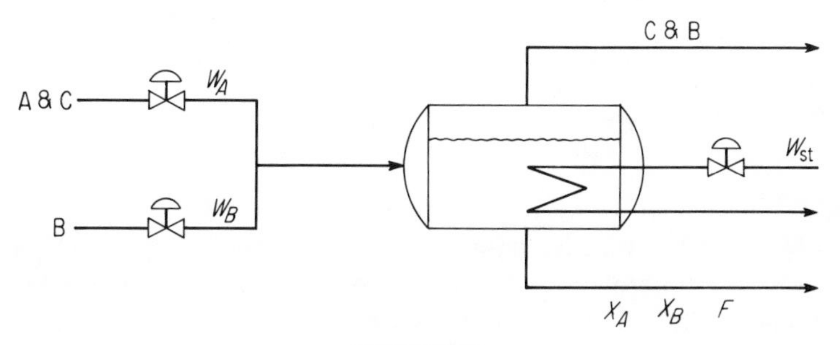

PROB. 8-7

Controlled variables

$$X_A, X_B, F$$

Manipulated variables

$$W_A, W_B, W_{st}$$

Total material balance

$$W_A + W_B - V = F \tag{1}$$

Material balance on A

$$X_A = \frac{W_A X_{A_{\text{in}}}}{W_A + W_B - V} \tag{2}$$

Material balance on B

$$X_B = \frac{W_B - V X_{B_o}}{W_A + W_B - V} \tag{3}$$

Heat balance on liquid

$$-(W_A + W_B) C_{p_{\text{ave}}} (T - T_{\text{in}}) + W_{st} \lambda_{st} + V \; [C_p(T - T_R) - \lambda_{CB}] = 0$$

Numerical values at steady state

F = 100 lb/hr
X_A = 0.7
X_B = 0.2
W'_A = 100(0.7) = 70
V = 20 lb/hr
W_A = 70 + 20 + 10 = 100
W_B = 20 lb/hr
λ_{st} = 825.1 Btu/lb
W_{st} = 50 lb/hr
C_p = 1 Btu/lb° F
$C_{p_{\text{ave}}}$ = 0.6 Btu/lb° F
T_{in} = 77° F
T_R = 32° F
λ_{CB} = 1000 Btu/lb
$X_{A_{\text{in}}}$ = 0.7
X_{B_o} = 0.1

Develop the relative gain matrix and suggest the proper pairing. Derive the steady-state decoupler with unit gains on the diagonal when proper pairing is used.

8-8. Consider the control system for the evaporator shown.

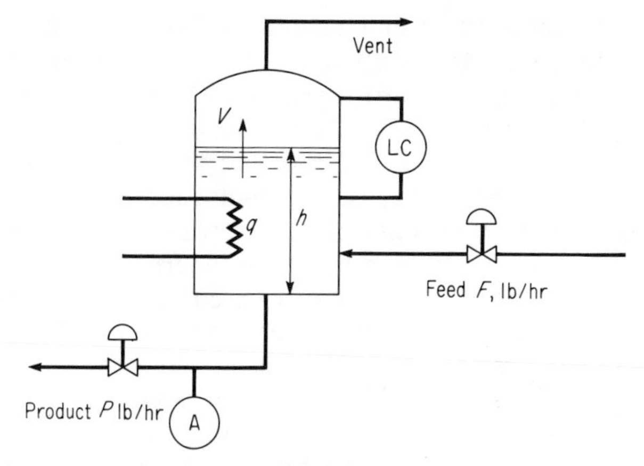

PROB. 8-8

Feed: $A + B$, would like to concentrate A by vaporizing B
Controlled variables: h, X_A
Manipulative variables: F, P
where h = level, ft
X_A = mass fraction (lb of A/total lb)

Overall mass balance

$$F - P - V = A\,\rho\,\frac{dh}{dt}$$

Mass balance on A (assume no A in V; well mixed)

$$FX_{Ai} - PX_A = 0$$

Numerical values

$$F = 100 \text{ lb/hr}$$
$$X_{Ai} = 0.5$$
$$V = 30 \text{ lb/hr}$$
$$P = 70 \text{ lb/hr}$$

Develop the relative gain matrix and suggest the proper pairing. Derive the steady-state decoupler with unit gains on the diagonal when proper pairing is used.

8-9. In the paper "Interaction Analysis in Control System Design" by Nisenfeld and Schultz (*Proceedings* of the 25th Annual ISA Conference and Exhibit, Philadelphia, October 1970, paper 562), an extension of the relative gain concept called the interaction index is presented. Apply this concept to the system in Exercise 8-4.

8-10. The simple model of the Fourdrinier paper machine illustrated in the accompanying illustration was used by Dahlin to design a decoupled control system (E. B., Dahlin et al., "Designing and Tuning Digital Controllers," Part II, *Instruments and Control Systems*, Vol. 41, No. 7 (July 1968), pp. 87–91.)

(a) Design a decoupler. Note that the straightforward technique suggested in Sec. 8-3 yields a positive dead time in one of the decoupler elements, which is physically unrealizable. A dead time could be incorporated into the D_{11} element of the decoupler to circumvent this problem, but would seriously degrade the performance of the moisture control system. For this reason, the positive dead time is generally simply deleted, which results in imperfect decoupling. However, the moisture control loop is not degraded. Note that this problem does not exist if K_{21} is essentially zero, which is true for most paper machines.

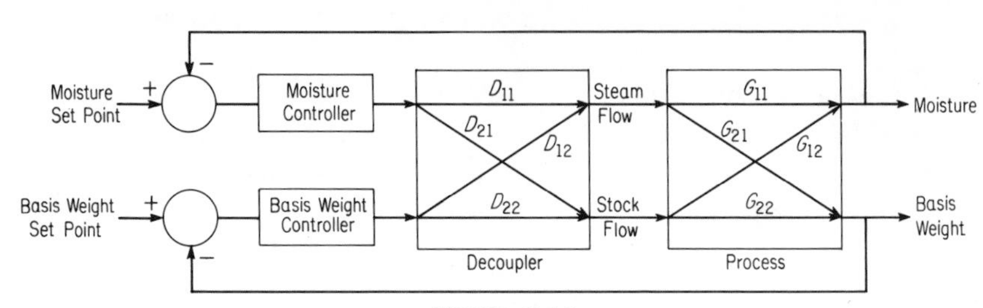

PROB. 8-10

$$G_{11} = \frac{K_{11}e^{-\theta_1 s}}{\tau_1 s + 1} \qquad G_{12} = \frac{K_{12}e^{-\theta_2 s}}{\tau_2 s + 1}, \quad \theta_2 > \theta_1$$

$$G_{21} = \frac{K_{21}e^{-\theta_1 s}}{\tau_1 s + 1} \qquad G_{22} = \frac{K_{22}e^{-\theta_2 s}}{\tau_2 s + 1}, \quad \theta_2 > \theta_1$$

(b) Design the individual controllers using Dahlin's techniques discussed in Chapter 6.

8-11. In Sec. 8-5 it was indicated that dead time compensation was inherently incorporated when the z-transform design techniques were used. To illustrate this, consider designing a control system for the process $\dfrac{K_1 e^{-\theta s}}{s}$.

(a) Design a controller using Dahlin's technique discussed in Chapter 6.

(b) Design a control system by first incorporating dead time compensation followed by design of the controller using Dahlin's technique. In this case, Dahlin's technique is applied to the dead-time-free system, namely K_1/s.

The two resulting control systems can be shown to be identical.

Index

DATE DUE

Summer Quarter '79			
[illegible] 7 '86			
OC 28 '89			
JA 13 '92			